FOOD SCIENCE AND TECHNOLOGY

FOOD FRAUD AND ADULTERATED INGREDIENTS

BACKGROUND, ISSUES, AND FEDERAL ACTION

FOOD SCIENCE AND TECHNOLOGY

Additional books in this series can be found on Nova's website under the Series tab.

Additional e-books in this series can be found on Nova's website under the e-book tab.

FOOD SCIENCE AND TECHNOLOGY

FOOD FRAUD AND ADULTERATED INGREDIENTS

BACKGROUND, ISSUES, AND FEDERAL ACTION

DARRELL T. BRADEN
EDITOR

New York

For permission to use material from this book please contact us:
Telephone 631-231-7269; Fax 631-231-8175
Web Site: http://www.novapublishers.com

Additional color graphics may be available in the e-book version of this book.

Library of Congress Cataloging-in-Publication Data

ISBN: 978-1-63117-730-9

Published by Nova Science Publishers, Inc. †New York

CONTENTS

PREFACE

Food fraud, or the act of defrauding buyers of food or ingredients for economic gain—whether they be consumers or food manufacturers, retailers, and importers—has vexed the food industry throughout history. Some of the earliest reported cases of food fraud, dating back thousands of years, involved olive oil, tea, wine, and spices. These products continue to be associated with fraud, along with some other foods. Although the vast majority of fraud incidents do not pose a public health risk, some cases have resulted in actual or potential public health risks. This book provides an overview of issues pertaining to food fraud and "economically motivated adulteration" or EMA, a category within food fraud. The book also examines the approaches that FDA uses to detect and prevent economic adulteration of food and medical products and the challenges FDA faces in detecting and preventing economic adulteration and views of stakeholders on options for FDA to enhance its efforts to address economic adulteration.

Chapter 1 – Food fraud, or the act of defrauding buyers of food or ingredients for economic gain—whether they be consumers or food manufacturers, retailers, and importers—has vexed the food industry throughout history. Some of the earliest reported cases of food fraud, dating back thousands of years, involved olive oil, tea, wine, and spices. These products continue to be associated with fraud, along with some other foods. Although the vast majority of fraud incidents do not pose a public health risk, some cases have resulted in actual or potential public health risks. Perhaps the most high-profile case has involved the addition of melamine to high-protein feed and milk-based products to artificially inflate protein values in products that may have been diluted. In 2007, pet food adulterated with melamine reportedly killed a large number of dogs and cats in the United States, followed by reports that melamine-contaminated baby formula had sickened thousands of Chinese children. Fraud was also a motive behind Peanut Corporation of America's actions in connection with the *Salmonella* outbreak in 2009, which killed 9 people and sickened 700. Reports also indicate that fish and seafood fraud is widespread, consisting mostly of a lower-valued species, which may be associated with some types of food poisoning or allergens, mislabeled as a higher-value species. Other types of foods associated with fraud include honey, meat and grain-based foods, fruit juices, organic foods, coffee, and some highly processed foods. It is not known conclusively how widespread food fraud is in the United States or worldwide. In part, this is because those who commit food fraud want to avoid detection and do not necessarily intend to cause physical harm. Most incidents go undetected since they usually do not result in a food safety risk and consumers often do not notice a quality problem. Although the full scale of food fraud is not known, the number of documented incidents may be a small fraction of the true number of incidents. The Grocery Manufacturers Association estimates that fraud may cost the global food industry between \$10 billion and \$15 billion per year, affecting approximately 10% of all commercially sold food products. Fraud resulting in a food safety or public health risk event could have significant financial or public relations consequences for a food industry or company. There is no statutory definition of food fraud or "economically motivated adulteration" (EMA) of foods or food ingredients in the United States. However, as part of a 2009 public meeting, the Food and Drug Administration (FDA) adopted a working definition, defining EMA as the "fraudulent, intentional substitution or addition of a substance in a product for the purpose of increasing the apparent value of the product or reducing the cost of its production, i.e., for economic gain." Efforts are ongoing to compile and capture current and historical data on food fraud and EMA incidents through the creation of databases and repositories. Over the years, Congress has introduced a number of bills intended to address concerns about food fraud for a particular food or food ingredient. Such legislation has not addressed food fraud in a comprehensive manner. However, although no single federal agency or U.S. law directly addresses food fraud, a number of existing laws and statutes already provide the authority for various federal agencies to address fraud. Currently, food fraud is broadly addressed through various food safety, food defense, and food quality authorities as well as border protection and import authorities across a number of federal agencies. FDA and the U.S. Department of Agriculture are the principle agencies that are working to protect the food supply from food safety risks—both unintentionally and intentionally introduced contamination—in conjunction with border protection and enforcement activities by the U.S. Department of Homeland Security. Other agencies also play a role.

Chapter 2 – Media attention has focused on recent incidents of fraud relating to seafood—restaurants knowingly serving lower-priced fish than identified on menus, lower-priced species marketed commercially as higher-priced species, packaged weights of seafood being less than labeled weights, and extra water being added to seafood to increase total product weight—raising public concern. Most seafood fraud is based on supplying the consumer with something different from and inferior to the product expected. In some instances, such practices may not be intentional, and the extent of intentional fraud is not well documented. These occurrences have the potential to erode consumer confidence in seafood generally, potentially lowering Americans' consumption of seafood. On a broader scale, seafood fraud is recognized as one component of international concerns related to illegal, unreported, and unregulated (IUU) fish harvests. Congress is facing questions of whether U.S. law applicable to fraudulent seafood sales and marketing is clear and enforceable, whether agency enforcement efforts targeting seafood fraud are adequate, and whether the penalties for seafood fraud are a deterrent. The primary federal law that addresses mislabeling is the Federal Food, Drug, and Cosmetic Act of 1938 (FFDCA; 21 U.S.C. §§ 301 et seq.), which is administered by the Food and Drug Administration (FDA). The FFDCA, as amended, gives FDA authority over most food regulation and includes (1) definitions elaborating on the concepts of adulteration and misbranding; (2) provisions for control over all labeling of foods in interstate commerce; (3) detailed regulation of safety and wholesomeness of foods; and (4) enforcement remedies available to the agency. In addition, the Fair Packaging and Labeling Act (FPLA; 15 U.S.C. §§ 1451 et seq.) requires that consumers of packaged commodities be provided with accurate information as to its contents. Food products falling within the scope of the FFDCA that are introduced into interstate commerce in violation of the FPLA and its regulations are deemed to be misbranded within the meaning of the FFDCA. The Chemical Services Group of the National Marine Fisheries Service (National Oceanic and Atmospheric Administration, Department of Commerce) at the National Seafood Inspection Laboratory in Pascagoula, MS, provides analytical testing to identify fish species. Applicable law is discussed in detail below.

Chapter 3 – In recent years, the United States experienced public health crises suspected to have been caused by the deliberate substitution or addition of harmful ingredients in food and drugs—specifically melamine in pet food and oversulfated chondroitin sulfate in the blood thinner heparin. These ingredients were evidently added to increase the apparent value of these products or reduce their production costs, an activity GAO refers to as economic adulteration. The Food and Drug Administration (FDA), an agency within the Department of Health and Human Services (HHS), has responsibility for protecting public health by ensuring the safety of a wide range of products that are vulnerable to economic adulteration. This report examines (1) the approaches that FDA uses to detect and prevent economic adulteration of food and medical products and (2) the challenges FDA faces in detecting and preventing economic adulteration and views of stakeholders on options for FDA to enhance its efforts to address economic adulteration. GAO reviewed FDA documents and interviewed FDA officials and stakeholders from academia and industry, among others.

In: Food Fraud and Adulterated Ingredients ISBN: 978-1-63117-730-9
Editor: Darrell T. Braden

Chapter 1

FOOD FRAUD AND "ECONOMICALLY MOTIVATED ADULTERATION" OF FOOD AND FOOD INGREDIENTS*

Renée Johnson

SUMMARY

Food fraud, or the act of defrauding buyers of food or ingredients for economic gain—whether they be consumers or food manufacturers, retailers, and importers—has vexed the food industry throughout history. Some of the earliest reported cases of food fraud, dating back thousands of years, involved olive oil, tea, wine, and spices. These products continue to be associated with fraud, along with some other foods. Although the vast majority of fraud incidents do not pose a public health risk, some cases have resulted in actual or potential public health risks. Perhaps the most high-profile case has involved the addition of melamine to high-protein feed and milk-based products to artificially inflate protein values in products that may have been diluted. In 2007, pet food adulterated with melamine reportedly killed a large number of dogs and cats in the United States, followed by reports that melamine-contaminated baby formula had sickened thousands of Chinese children. Fraud was also a motive behind Peanut Corporation of America's actions in connection with the *Salmonella* outbreak in 2009, which killed 9 people and sickened

* This is an edited, reformatted and augmented version of Congressional Research Service Publication, No. R43358, dated January 10, 2014.

700. Reports also indicate that fish and seafood fraud is widespread, consisting mostly of a lower-valued species, which may be associated with some types of food poisoning or allergens, mislabeled as a higher-value species. Other types of foods associated with fraud include honey, meat and grain-based foods, fruit juices, organic foods, coffee, and some highly processed foods.

It is not known conclusively how widespread food fraud is in the United States or worldwide. In part, this is because those who commit food fraud want to avoid detection and do not necessarily intend to cause physical harm. Most incidents go undetected since they usually do not result in a food safety risk and consumers often do not notice a quality problem. Although the full scale of food fraud is not known, the number of documented incidents may be a small fraction of the true number of incidents. The Grocery Manufacturers Association estimates that fraud may cost the global food industry between $10 billion and $15 billion per year, affecting approximately 10% of all commercially sold food products. Fraud resulting in a food safety or public health risk event could have significant financial or public relations consequences for a food industry or company.

There is no statutory definition of food fraud or “economically motivated adulteration” (EMA) of foods or food ingredients in the United States. However, as part of a 2009 public meeting, the Food and Drug Administration (FDA) adopted a working definition, defining EMA as the “fraudulent, intentional substitution or addition of a substance in a product for the purpose of increasing the apparent value of the product or reducing the cost of its production, i.e., for economic gain.” Efforts are ongoing to compile and capture current and historical data on food fraud and EMA incidents through the creation of databases and repositories.

Over the years, Congress has introduced a number of bills intended to address concerns about food fraud for a particular food or food ingredient. Such legislation has not addressed food fraud in a comprehensive manner. However, although no single federal agency or U.S. law directly addresses food fraud, a number of existing laws and statutes already provide the authority for various federal agencies to address fraud. Currently, food fraud is broadly addressed through various food safety, food defense, and food quality authorities as well as border protection and import authorities across a number of federal agencies. FDA and the U.S. Department of Agriculture are the principle agencies that are working to protect the food supply from food safety risks—both unintentionally and intentionally introduced contamination—in conjunction with border protection and enforcement activities by the U.S. Department of Homeland Security. Other agencies also play a role.

BACKGROUND

Food fraud, or the act of defrauding buyers of food and food ingredients for economic gain— whether they be consumers or food manufacturers, retailers, and importers—has vexed the food industry throughout history. Some of the earliest reported cases of food fraud, dating back thousands of years, involved olive oil, wine, spices, and tea.[1] These same products continue to be associated with fraud, along with a range of other products. Overall, foods and food ingredients commonly associated with food fraud include olive oil, fish, honey, milk and dairy products, meat products, grain-based foods, fruit juices, wine and alcoholic beverages, organic foods, spices, coffee, and tea, and some highly processed foods. It is not known conclusively how widespread food fraud is in the United States or worldwide. In part, this is because those who commit fraud do not intend to cause physical harm and want to avoid detection. Most incidents go undetected since they usually do not result in a food safety risk and consumers often do not notice a quality problem. Moreover, as the motivation to commit fraud is illicit monetary gain, the type of food that might be or become adulterated is a secondary consideration (i.e., it could be any type of food or food ingredient); rather, it is the opportunity or feasibility of committing fraud that generally triggers the fraud.[2]

Although the vast majority of food fraud incidents do not pose a public health risk, there have been fraud cases that have resulted in actual or potential public health risks. Perhaps the most widely cited, high-profile cases have involved the addition of melamine to high-protein feed and milk-based products to artificially inflate protein values in products that may have been diluted. For example, in 2007, evidence emerged that adulterated pet food ingredients from China had caused the deaths of a large number of dogs and cats in the United States.[3] This was followed by reports that melamine-contaminated baby formula had sickened an estimated 300,000 Chinese children, killing a reported 6 infants.[4] Evidence now suggests that safety risks associated with melamine-tainted feed date back to 2003, and that melamine was first reported to be added to artificially increase protein content in feed as far back as 1982.[5]

Reports also indicate fish and seafood fraud may be widespread in some markets, consisting mostly of the mislabeling or substitution of a higher-valued species with something different from and inferior to the expected species, possibly with a fish species which could be associated with some types of food poisoning or exposure to certain allergens.[6] Similarly, substitution of olive oil with other types of seed, legume, or nut oils could have unintended consequences, if consumed by those with certain food allergies.

Some cases might not initially appear to involve intentional adulteration, except on closer examination. Charges of fraud were part of the federal criminal indictment charging former officials of the Peanut Corporation of America with numerous offenses in connection with the *Salmonella* outbreak in 2009—which killed 9 people and sickened 700—since company officials were found to have sold and distributed product known to be contaminated.[7] That case resulted in one of the largest product recalls in U.S. history, including 3,912 products that contain peanut butter and peanut paste ingredients, such as cookies, crackers, cereal, candy, ice cream, pet treats, and other foods, which were manufactured by more than 200 companies.[8]

Risks from other types of fraudulent foods are not as well-documented and may be less immediate or may never be known. Generally, only those who have knowledge of the fraud are those who commit the fraud. In some cases, independent tests might uncover fraud. For example, FDA testing upon import has documented the presence of unapproved chemicals in honey, which have triggered import alerts.[9] Reports indicate that honey from China may contain certain unapproved antibiotics or other agricultural chemicals.[10] Reports also indicate that some fruit juices may be made from or diluted with juice from rotten fruit and may contain toxic mold.[11] In recent years, fraud involving the addition of certain food processing aids (known as "clouding agents") has raised concern given the potential public health risks and reportedly increased use in certain highly processed foods.[12] Some fraud cases might only be uncovered following an investigation in the wake of a public health event, such as when pet food adulterated with melamine caused the deaths of dogs and cats in the United States. Some fraud cases, however, might never be discovered even though they could contribute to chronic long-term health consequences.

Other food fraud concerns might not result in a public health or food safety crisis, but instead deprive the food buyer of the product they think they are getting. Most food fraud cases involve the substitution of a high-value product with a less expensive or lower quality alternative. Such cases include cheaper products mislabeled as extra virgin olive oil from Italy, wild Alaskan salmon, caviar, and pomegranate juice or juices from other "super" fruit. In another example, in Europe in early 2013, it was reported that products labeled as containing beef were found to actually contain 80%-100% horsemeat, which the meat supplier had knowingly failed to report to local authorities.[13]

Although the horsemeat incident ultimately did not result in public health consequences, initial concerns about potential health risks due to, for example, phenylbutazone[14] resulted in a substantial expenditure of public resources over the course of the investigation. Such cases also erode consumer confidence and may cause other concerns for a variety of societal or cultural reasons.

The Grocery Manufacturers Association (GMA) estimates that fraud may cost the global food industry between $10 billion and $15 billion per year,[15] affecting approximately 10% of all commercially sold food products.[16] However, most researchers acknowledge that the full scale of food fraud "may be unknown or even possibly unknowable"[17] even though the number of documented incidents is "most likely a fraction of the true number of incidents, since the goal of adulteration for economic gain is not to be detected."[18] Compared to the trillions of dollars spent on food and food ingredients globally each year, however, "the prevalence of food fraud is ultimately very low."[19] Fraud resulting in a food safety or public health risk event, however, could have significant financial or public relations consequences for a food industry or company. GMA estimates that fraud costs food businesses in terms of lost sales, estimated between 2% and 15% of annual revenues, as well as possible bankruptcies if adverse public health consequences occur.

The text box on the next page describes some of the types of foods and food ingredients associated with fraud based on available research and information.

Addressing food fraud concerns has become exceedingly complicated by rising U.S. imports and increased globalization of the world's food and agricultural supplies. Increasingly, multiple product ingredients and inputs are sourced from a range of countries, both in terms of individually sourced products and ingredients between individual food companies/importers as well as in terms of internally sourced products from foreign-owned entities within a larger multinational company (such as global sourcing between parent and subsidiary). It is difficult to detect and trace not only the source of unintentional contamination and related food safety concerns, but it is often more difficult to detect and trace-back instances of intentional product fraud, especially in highly processed foods with multiple ingredients and inputs from multiple suppliers.

This report provides an overview of issues pertaining to food fraud and "economically motivated adulteration" or EMA, a category within food fraud. First, the report provides general background information on food fraud and EMA, including how it is defined and the types of fraud, as well as how food fraud fits into the broader policy realm of food safety, food defense, and food quality. Second, the report provides available information about foods and ingredients with reported cases of fraud from two databases: (1) the United States Pharmacopeial Convention (USP) Food Fraud Database and (2) the National Center for Food Protection and Defense (NCFPD) EMA Incident Database. Finally, the report describes previous and ongoing federal and congressional actions to address food fraud.

Leading Food Categories with Reported Cases of Food Fraud

Olive Oil. Olive oil is often substituted with a lower cost alternative, whether it is regular olive oil instead of higher-priced extra virgin olive oil or a less expensive variety from Greece or Turkey, instead of from Italy as the label claims. In such cases the fraud was associated with efforts to defraud the European Union's farm support program, which subsidizes olive oil, as part of the Common Agricultural Policy (CAP). In some cases an alternate seed or nut oil may be sold as or thinned out with hazelnut, soybean, corn, peanut, sunflower, safflower, walnut, vegetable, canola, or palm oil, and in one case, lard. Some combinations contained no olive oil. The use of nut or legume oils could pose a problem for those with certain food allergies. In rare cases, non-food-grade oil may be added, such as rapeseed.

Fish and Seafood. Some higher-value fish and seafood are replaced with cheaper, more abundant fish. A report by Oceana found that fish samples purchased at grocery stores, restaurants, and sushi bars in major cities were often mislabeled, including red snapper (actually tilefish); white tuna and butterfish (actually escolar); wild Alaskan salmon (actually farmed Atlantic salmon); caviar (actually catfish roe); and monkfish (puffer fish). Other types of substitutions have involved halibut, sole, grouper, and striped bass. Some substitutions have involved fish or seafood associated with certain types of fish poisoning or allergens. Other substitutions are intended to evade import and other restrictions.

Milk and Milk-based Products. Milk from bovine cows has had milk from other types of animals, such as sheep, buffalo, and goats-antelopes, added to it, but also adulterated with reconstituted milk powder, urea, and rennet, among other products (oil, detergent, caustic soda, sugar, salt, and skim milk powder). Adulterated milk may also be watered down and then supplemented with melamine to artificially raise the apparent protein content and hide dilution.

Melamine, an organic base chemical, is widely used in plastics, adhesives, and other consumer products, and is known to pose a public health threat. Adulterated milk might also be added into infant formula and other milk-based products. Baby formula is a common target for retail theft, often by tampering with the sell-by codes to move expired product.

Honey, Maple Syrup, and Other Natural Sweeteners. Honey might have added sugar syrup, corn syrup, fructose, glucose, high-fructose corn syrup, and beet sugar, without being disclosed on the label. Honey from a "non-authentic geographic origin" is also common, such as cases where honey from China is transshipped through another Asian country and falsely sold as honey from the second country—usually to avoid higher customs duties and tariffs that would be imposed on honey from China. Some of this honey might also contain unapproved antibiotics or other additives and heavy metals. Maple syrup is sometimes thinned out with sugar or corn syrup.

Fruit Juice. Juices might be watered down, or a more expensive juice (such as from pomegranates or other "super" fruit) might be cut with a cheaper juice (such as apple or grape juice). Some juice may be only water, dye, and sugary flavorings, although fruit is the listed ingredient on the label.

Orange juice has been shown to sometimes contain added unlisted lemon juice, mandarin juice, grapefruit juice, high fructose corn syrup, paprika extract, and beet sugar. Apple juice has been shown to have added unlisted grape juice, high fructose corn syrup, pear juice, pineapple juice, raisin sweetener, fig juice, fructose, and malic acid.

Coffee and Tea. Ground coffee might be cut with leaves and twigs, as well as roasted corn, ground roasted barley, and roasted ground parchment. Instant coffee may include chicory, cereals, caramel, more parchment, starch, malt, and figs. Tea may contain leaves from other plants, color additives, and colored saw dust.

Spices. Saffron is the world's most expensive spice, and has been found to have added glycerin, sandalwood dust, tartrazine (a yellow dye), barium sulfate, and borax. Ground black pepper has been shown to have added starch, papaya seeds, buckwheat, flour, twigs, and millet. Vanilla extract, turmeric, star anise, paprika, and chili powder are other spices prone to fraud. Sudan red dyes have been used to color paprika, chili powders, and curries, but are also known carcinogens and are banned for use in foods.

Organic Foods and Products. Using fraudulent certification to market, label, or sell non-organic (conventionally produced) agricultural products as USDA-certified "organic" is a violation of U.S. law and federal National Organic Program (NOP) regulations. Products fraudulently labeled as "organic" have been detected by USDA for a range of foods and food ingredients from both domestic and international suppliers.

Clouding agents. "Clouding agents" or food processing aids "to enhance the appeal or utility of a food or food component," such as palm oil and other allowed food ingredients, are often used in fruit juices, jams, and other foods. Of particular concern is the fraudulent replacement or addition of the plasticizer Di(2-ethylhexyl) phthalate (DEHP) and other related phthalates, as a substitute for other ingredients. DEHP may also be used in food contact materials, such as seals and packaging. DEHP is associated with public health risks, including cancer and reproductive concerns.

Source: CRS compilation from information reported by USP, Michigan State University, NCFPD and researchers at the University of Minnesota, Oceana, Consumers Union, *Food Chemical News*, and the Rodale Institute. Unless otherwise indicated, "adulteration" and "misbranding" of foods is prohibited under various FDA and USDA laws.

Existing Definitions

There is no statutory definition of food fraud or "economically motivated adulteration" or EMA of foods or food ingredients, which is generally considered a subset of food fraud. However, the U.S. Food and Drug Administration (FDA) adopted a "working definition" for an April 2009 public meeting to raise awareness and solicit public input regarding economically motivated adulteration of FDA-regulated products. For the purposes of the workshop, FDA defined "economically motivated adulteration" as the:[20]

> fraudulent, intentional substitution or addition of a substance in a product for the purpose of increasing the apparent value of the product or reducing the cost of its production, i.e., for economic gain. EMA includes dilution of products with increased quantities of an already-present substance (e.g., increasing inactive ingredients of a drug with a resulting reduction in strength of the finished product, or watering down of juice) to the extent that such dilution poses a

> known or possible health risk to consumers, as well as the addition or substitution of substances in order to mask dilution.

Other countries generally also do not have established legal definitions of food fraud. A recent European Union (EU) report on food fraud states the EU laws do not provide for a "generally acknowledged definition of food fraud" despite an extensive legislative framework focused on food safety.[21] The only general guideline is found in EU regulations requiring that food labeling, advertising, presentation, and packaging "shall not mislead consumers."[22] However, these requirements reportedly vary among EU member states and food fraud in Europe remains largely undetected, similar to that in the United States. The United Kingdom's Food Standards Agency (FSA) describes "food fraud" as the deliberate placement on the market, for financial gain, with the intention of deceiving the consumer, covering two main types of fraud.[23] These include the sale of food which is unfit and potentially harmful[24] as well as the deliberate misdescription of food, such as products substituted with a cheaper alternative.[25]

Researchers and industry groups actively working in this area have myriad definitions of food fraud and EMA.

Broadly speaking, food fraud is a type of product fraud. According to food safety researchers at Michigan State University's (MSU's) Food Fraud Initiative:[26]

> Food fraud is a collective term used to encompass the deliberate and intentional substitution, addition, tampering, or misrepresentation of food, food ingredients, or food packaging; or false or misleading statements made about a product for economic gain.

The United States Pharmacopeial Convention (USP) states:[27]

> Food fraud in the context of food ingredients refers to the fraudulent addition of non-authentic substances or removal or replacement of authentic substances without the purchaser's knowledge for economic gain of the seller. It is also referred to as economic adulteration, economically motivated adulteration, intentional adulteration, or food counterfeiting.

In this context, "adulterant" is defined as "the undesirable substance" or "fraudulently added material" in a fraudulent food or food ingredient. In the case of food fraud, the adulterants used often are unconventional and designed to avoid detection through routine inspection.[28]

According to researchers at the National Center for Food Protection and Defense (NCFPD) at the University of Minnesota:[29]

> Economically motivated adulteration (EMA) is the intentional sale of substandard food or food products for the purpose of economic gain. Common types of EMA include intentional substitution of an authentic ingredient with a cheaper product, dilution with water or other substances, flavor or color enhancement using illicit or unapproved substances, and substitution of one species with another.

The Grocery Manufacturers Association (GMA) provides the following definitions of "economic adulteration" as part of its report on consumer product fraud in the food, beverage, and consumer product industry:[30]

> Economic adulteration is defined as the intentional fraudulent modification of a finished product or ingredient for economic gain through the following methods: unapproved enhancements, dilution with a lesser-value ingredient, concealment of damage or contamination, mislabeling of a product or ingredient, substitution of a lesser-value ingredient or failing to disclose required product information.

These definitions broadly reference three types of fraud, as defined by USP and other university researchers, namely:[31]

- **complete or partial replacement** of a food ingredient or valuable authentic constituent with a less expensive substitute (or alternative animal species in the case of some meat and fish);
- **addition** of small amounts of a non-authentic substance to mask inferior quality ingredient (or excess packing ingredients, including water and ice); and
- **removal or intentional omission** of an authentic and valuable constituent in a food product or food ingredient.

Also included are false claims and non-declarations based either on geographic, species, botanical, varietal origin, and production processes in order to provide economic benefit through the substitution of a particular food or food ingredient with a lower priced or a lower quality ingredient. Fraud may also be motivated in cases where a particular food or food ingredient is in short supply, triggering the motivation to substitute one input for another. In other cases false declarations of origin are intended to evade taxes and tariffs in importing countries. Food fraud also includes smuggling, tampering, and stolen goods. The text box on the next page provides more detailed information.

Risks to Food Protection

Figure 1 provides a matrix of food protection risk and illustrates how food fraud fits into the broader policy realm of food safety, food defense, and food quality.[32] Although these concepts may not always fit neatly into these categories and there may be overlap across these categories, this matrix provides a useful, and widely cited, framework for differentiating among these food protection areas. In general, food fraud and food defense (agro-terrorism) are both intentional. Food fraud is always economically motivated. Motivation in the case of food defense includes the intention to inflict public harm or threaten consumers, and is considered to be food fraud.[33] Food fraud usually is perpetrated by actors normally involved in the food chain that have regular access to the food product (e.g., manufacturers or distributors). Problems with food defense usually are perpetrated by outsiders, including terrorists, who do not normally have access to the food product. NCFPD reports that most incidents of food fraud generally do not result in public health harm; however, sometimes perpetrators of food fraud “make mistakes and unintended health consequences result.”[34] Table 1 contrasts the different types of food-related risks, and provides examples, causes and motivations, primary and secondary effects, and also types of public health risks.

Types of Food Fraud

(1) Replacement: Complete or partial replacement of a food ingredient or valuable authentic constituent with a less expensive substitute, usually achieved through the *addition, dilution, or extension* of an authentic ingredient with an adulterant or mixture of adulterants. Examples include addition of melamine to milk to artificially increase apparent protein contents; addition of water and citric acid to lemon juice to fraudulently increase the titratable acidity of the final juice product; and overtreating frozen fish with ice (adds extra water, potentially unsanitary).

Included are false claims and non-declarations:

- False declaration of geographic, species, botanical, or varietal origin (e.g., substitution of less expensive cow's milk for sheep's or goat's milk; substitution of common wheat for durum wheat; substitution of Greek or Turkish olive oil labeled as Italian olive oil; substitution of synthetically produced vanillin for botanically derived natural vanillin); farmed salmon sold as wild; and the sale of meat from animals that have been stolen and/or illegally slaughtered, such as wild game animals that may have been poached.
- False declaration of origin to evade taxes/tariffs (e.g., U.S. imports of catfish from Vietnam labeled as grouper to avoid anti-dumping duties; transshipment of Chinese products, such as shrimp and honey, through other Asian nations to avoid anti-dumping duties or to mask the true origin of potentially unsafe product).
- False declaration of production process (e.g., fraudulent labeling of a synthetically derived flavor chemical as being "naturally" derived; food falsely labeled or certified as organically produced); or poor quality filtered honey or a honey substitute product labeled as raw honey, which reportedly has medicinal properties.

Such fraudulent practices are generally intended to provide economic benefit through the substitution of a particular food or food ingredient with a lower priced or a lower quality ingredient.

In some cases, short supply of a particular food or food ingredient might trigger the motivation to substitute one input for another, for example, when horse meat was found in prepared meals as a result of a reduced European supply of beef and other meat products.

(2) Addition: Addition of small amounts of a non-authentic substance to mask inferior quality ingredient. An example is the addition of a color additive (such as Sudan Red dyes) to enhance the color of poor quality paprika.

(3) Removal: Removal or intentional omission of an authentic and valuable constituent without the purchasers' knowledge.

One example is the removal of non-polar constituents from paprika (e.g., lipids and flavor compounds) to produce paprika-derived flavoring extracts, or "defatted" paprika, which lacks valuable flavoring compounds, as normal paprika. Another example is poor quality honey that has been filtered to remove pollen or other residue from the beehive, in order to make it difficult to determine the honey's botanical and geographic origin or to circumvent the ability to trace and identify the actual source of the honey.

Source: J.C. Moore, J. Spink, and M. Lipp, "Development and Application of a Database of Food Ingredient Fraud and Economically Motivated Adulteration from 1980 to 2010," *Journal of Food Science*, Vol. 77, Nr. 4, 2012; and USP, Food Fraud Database, Glossary of Terms, http://www.foodfraud.org /glossary-terms.

Note: "Transshipment" refers to the transfer of a shipment from one carrier or vessel to another during transit, usually to hide the identity of the port or country of origin.

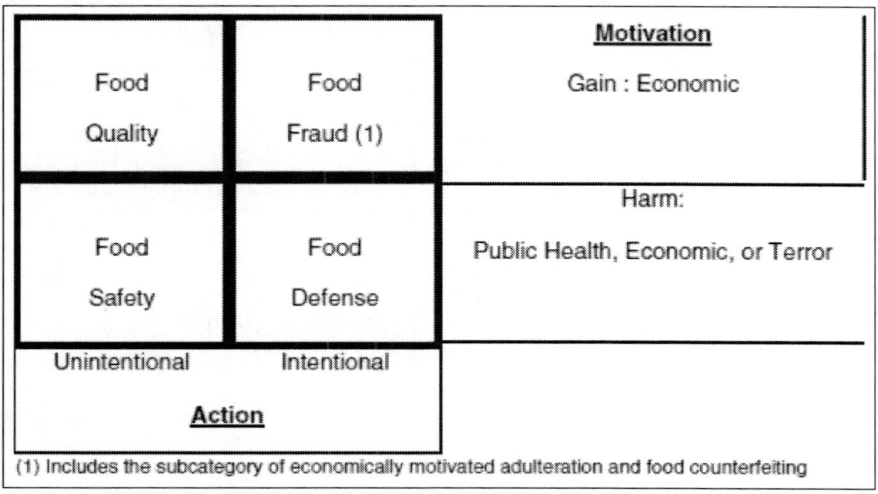

Source: J. Spink and D.C. Moyer, "Defining the Public Health Threat of Food Fraud," *Journal of Food Science*, 2011.

Figure 1. Food Protection Risk Matrix.

In contrast, problems with food safety and food quality are unintentional. Both food quality and food safety incidents may result in an economic impact to a particular food or food ingredient industry, for example, due to reduced purchases or brand equity, but also from product recalls, process controls, and liability in the case of food safety. Food safety concerns can also cause a threat to public health and create public fear. Adulteration will cause a loss of the consumer's trust in the food supply chain, and a loss in trust among regulators, industry, and U.S. trading partners, and may result in market and trade disruptions. Unintentional adulteration involving food quality concerns might also occur due to environmental factors, and packaging, storage, and distribution issues, among other factors.

Table 1. Food Protection Risk: Examples, Cause and Effects

Risk Type	Example	Cause and Motivation	Effect	Public Health Risk Type	Secondary Effect
Food Quality	Accidental bruising of fruit	Mishandling	Unsalable product or possible additional contamination	None, or possible food safety	Reduced product/brand equity or food safety incident
Food Fraud	Intentional adulteration of milk with melamine	Increased profit margins	Toxic poisonings	Food safety	Public fear and possible lower prices industry-wide
Food Safety	Unintentional contamination of raw vegetables with E. coli	Limited field protection and control during harvesting and processing	Illness and/or death	Food safety	Damaged industry, recall expense, and public fear
Food Defense	Intentional contamination of ground beef with nicotine	Revenge against the store/manager through injury to consumers	Nonlethal poisonings	Food defense	Adulterated product, damaged industry, recall expense, public fear

Source: J. Spink and D.C. Moyer, "Defining the Public Health Threat of Food Fraud," *Journal of Food Science*, 2011.

Available Data and Information Repositories

Review of Available Database Information

Efforts are ongoing to compile and capture current and historical data on food fraud and EMA incidents through the creation of databases and repositories. Although the information in these databases is not comprehensive and may contain certain limitations, it does represent the best information available and provides a first step to understand the scope and scale of food fraud as a way to further detect, combat, and prevent future fraud. The information contained in these databases is from the University of Minnesota as well as the United States Pharmacopeial Convention (USP), a long-standing scientific nonprofit organization. In addition, researchers from Michigan State University conducted further analysis of the USP databases, along with USP researchers. These databases provide a relatively new resource to examine food fraud incidents, and were first published starting in 2012 and 2013. Since then, information from these databases has been published in peer reviewed journal articles and presented at professional conferences by both academic researchers and industry leaders.[35] Nevertheless, because there is no single comprehensive surveillance system to detect food fraud in the United States or worldwide, it is not possible to know what portion of food fraud incidents may be captured by these databases, and whether or not the information in these databases represents the universe of potential food fraud incidents.

Information in these databases also has been widely cited by academic researchers, government officials, and the media to rank the leading foods and food ingredients that are most often associated with fraud. For example, information in the USP Food Fraud Database was used by the European Parliament to help identify the leading fraudulent foods, as part of its recent draft report calling for increased oversight of food fraud.[36] NDFPD's database, and preliminary analysis of information obtained from USP's database, was originally funded by the U.S. Government through funding from the Department of Homeland Security (DHS), FDA, and the U.S. Department of Agriculture (USDA), as well as funding from the private sector. USP's database and related activities are self-funded and provided in accordance with USP's mission to support public health, and are not supported by funding from any outside agency.

The two available databases are:

1) United States Pharmacopeial Convention (USP) Food Fraud Database, and
2) National Center for Food Protection and Defense (NCFPD) EMA Incident Database.

USP's database is open and publicly accessible; NCFPD's databases are accessible upon request.

Information in these databases is from available scholarly journal articles and industry analyses and lab tests (where available), as well as media reports, about foods and food ingredients that are vulnerable to fraudulent manipulation. These records are based on English-language scholarly and/or media sources. Both databases are largely global in scope and broadly represent fraud in foods and food ingredients sold commercially worldwide. Undoubtedly the information in these databases does not capture all cases of food fraud and EMA, or incidents that may occur in small-scale markets or on a non-commercial scale or incidents that may or may not be reported in various non-English speaking information outlets.

Since those who commit fraud actively seek to avoid detection, some food fraud and EMA cases might not be captured in either database. In fact, in some cases, records in the database may be more reflective of where research is being conducted or where resources have been dedicated in a concerted effort to root out fraud, rather than provide an exhaustive accounting of all fraud cases.

Differences between these two databases, including how they were compiled and the type of records they reference, among other things, may result in differences in how different foods and food ingredients may be ranked in terms of their susceptibility to food fraud and EMA.

These databases do not address fraud involving "dietary supplements."[37] Dietary supplements are a special category of food that includes finished products (e.g., a vitamin D tablet) that contain one or more "dietary ingredients" or are components of those finished products (e.g., vitamin D added to a food product such as breakfast cereal).[38] Like food fraud, fraud involving dietary supplements is a type of product fraud with documented concerns involving public health risks.[39] Additional research in this area is currently being considered, including development of a database of reported cases of dietary supplement fraud.[40]

USP Food Fraud Database

The United States Pharmacopeial Convention (USP) Food Fraud Database is a public database that catalogues available analytical methods to detect and identify problematic food ingredients, which in turn provides a repository for ingredient fraud reports.[41] The database is comprised of both "scholarly" and "media" reports from available food science scholarly sources (and undocumented industry analyses, where available) and mainstream English-language media. The database is not an "incidents" database where individual records have been further grouped by source and time period, but instead it catalogues reports involving detection methods and analyses of food fraud incidents.[42] As a result, information in the database may be more representative of foods that are the most researched, and not necessarily foods that are the most adulterated. The USP database is organized by food ingredient categories and identifies the type of adulterant reported for each documented record and broadly classifies each record by the type of fraud (e.g., addition, replacement, removal), but does not provide other information such as where the product was originally produced.

The initial version of the database was published in 2012 in the USP's 8th Edition of the *Food Chemicals Codex*[43] and was accompanied by initial analyses of the database records published in the April 2012 issue of the *Journal of Food Science,*[44] with the support of researchers at the MSU's Food Fraud Initiative. Initial funding for MSU's contribution to the manuscript was provided by the National Center for Food Protection and Defense (NCFPD) through DHS, with other funding from FDA and USDA.[45] The USP database was updated in 2013 and now covers the period from 1980 through 2012, and contains nearly 2,100 records.

NCFPD EMA Incident Database

The National Center for Food Protection and Defense (NCFPD) EMA Incident Database is another database, which differs in that it catalogs isolated EMA incidents and is available to authorized users upon request.[46] (NCFPD is a University of Minnesota-led consortium and has been a Homeland Security Center of Excellence since July 2004.)[47] As part of the database, an "incident" is defined as:[48]

> a documented, isolated occurrence of EMA in a single food product or group of associated food products occurring within a defined time frame and with a distinct group of perpetrators. Incidents that are difficult to isolate to a specific time frame and/or group of perpetrators, or with characteristics common to multiple perpetrators, are entered as a single incident.

For example, the melamine adulteration of infant formula in China in 2007-2008 is recorded as one incident. Individual records therefore have been further grouped by adulterant (e.g., melamine) and time period when the incident is estimated to have occurred. The database records also do not include information on issues that might be of concern but that remain undocumented in the public domain. The database tracks EMA incidents in food products since 1980. There were 1,054 records in the database in mid-

2012, covering 302 incidents.[49] The EMA Incident Database identifies the type of adulterant reported for each documented incident and classifies each by the type of adulteration (including substitution/dilution, unapproved additives, mislabeling, transshipment/origin masking, and port shopping),[50] and also provides information on where the product was originally produced based on an examination of the available information.

Initial analyses of the EMA Incidents Database records were published in the April 2013 issue of the *Journal of Food Protection*.[51] The database and related work to develop predictive models and case studies are intended to help characterize discrete EMA incidents to better understand the incentive behind the adulteration, including the adulterant used, how the adulteration was discovered, and how to detect and deter future incidents and protect the U.S. food supply from deliberate or intentional acts of contamination or tampering.[52] In addition to the EMA Incidents Database, NCFPD is also developing an EMA Susceptibility Database that includes evaluations of the monographs in the USP Food Chemicals Codex for susceptibility to EMA.[53] NCFPD is also working with DHS on other projects, including development of an assessment tool to determine and document the most critical food and agriculture infrastructure at the state level.[54]

Leading Reported Types of Fraud

These available databases provide information on the types of foods and food ingredients associated with food fraud. How these products and ingredients are ranked, however, may differ considerably among various reports. Specifically, overall product rankings of the leading reported fraudulent foods and food ingredients may differ depending on:

- database referenced (i.e., whether USP or NCFPD database);
- source or type of records referenced (i.e., in the USP database, ranking differs based on whether compiled from the "scholarly" or "media" or total records; in the NCFPD database, ranking is based on number of "incidents");
- time period (1980 through 2010 or 1980 through 2012); and
- organization compiling the data (i.e., whether conducted by USP or NCFPD, or whether conducted by an outside organization using data from one of the databases).

USP Food Fraud Database

The initial version of the United States Pharmacopeial Convention (USP) Food Fraud database published in 2012 covered both "scholarly" and "media" records from 1980 through 2010, and was accompanied with analysis conducted by researchers at MSU and USP. This first version of the database consisted of a total of 1,305 records covering 361 discrete food ingredients that were based on 660 references (Table 2).

Records in the USP database indicate that the leading reported types of fraud by specific ingredient among the database's scholarly records (1980-2010) were olive oil (16%), milk (14%), honey (7%), saffron (5%), orange juice (4%), coffee (3%), and apple juice (2%).[55] Figure 2 provides a consolidated breakdown by major food ingredient category. By major food ingredient, oils (24%), milk (14%), and spices (11%) account for nearly 50% of all reported cases.

Table 2. Scope of USP Food Fraud Database 2012 Publication and 2013 Update

Database Version	Scholarly Records	Media Records	Total
2012 Publication (1980-2010 data)			
Number of Records	1,054	251	1,305
Number of Ingredients	250	147	397
Number of References	575	85	660
2013 Update (mostly additional 2011-2012 data)			
Total Number of Records (since 1980)	1,648	449	2,097
Total Number of References (since 1980)	742	182	924

Source: As reported in J.C. Moore, J. Spink, and M. Lipp, "Development and Application of a Database of Food Ingredient Fraud and Economically Motivated Adulteration from 1980 to 2010," *Journal of Food Science*, Vol. 77, Nr. 4, 2012; and A.G. Ebert, "The Food Chemicals Codex EMA Activities: The Food Fraud Database – What's Next?" Presentation at the USP Workshop of Economically Motivated Adulteration of Food Ingredients and Dietary Supplements, September 26–27, 2013.

Records in the USP database indicate that the leading reported types of fraud by specific ingredient among the database's media records (1980-2010) were fish (9%), honey (6%), olive oil (4%), chili powder (4%), milk (3%), black pepper (3%), and caviar (2%). Figure 3 provides a consolidated breakdown by major food ingredient category whereby natural flavoring complexes[56] (30%) and spices (19%) account for nearly 50% of all reported cases. The 2013 updates to the USP Food Fraud Database added another 792 records from 264 additional references to the database, consisting mostly of new information published in 2011 and 2012. This raised the total number to 2,097 records, based on 924 references (Table 2).

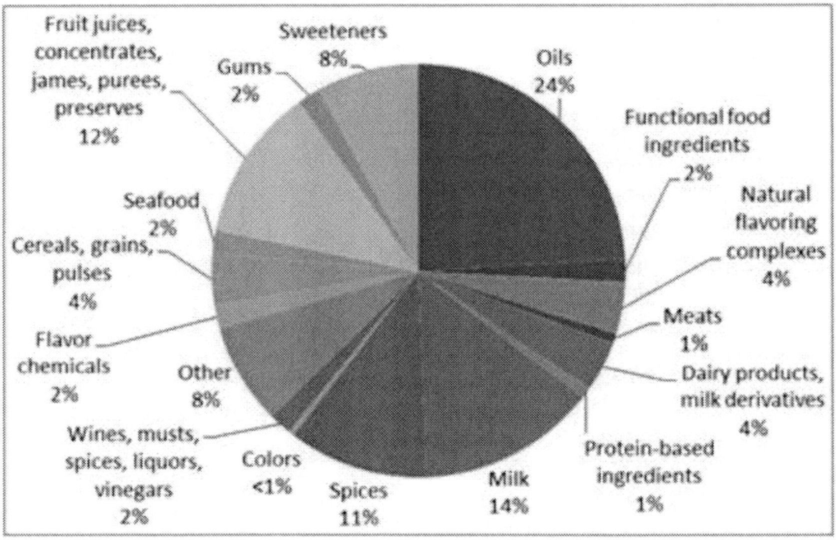

Source: USP Food Fraud Database (1980-2010), http://www.foodfraud.org/. As reported in J.C. Moore, J. Spink, and M. Lipp, "Development and Application of a Database of Food Ingredient Fraud and Economically Motivated Adulteration from 1980 to 2010," *Journal of Food Science*, Vol. 77, Nr. 4, 2012.

Notes: 1980-2010 data: records (1,054); ingredients (250); and references (575).

Figure 2. Leading Reported Types of Fraud, USP Scholarly Records (1980-2010) By Food Ingredient Category, USP Food Fraud Database.

Figure 4 provides a breakdown by major food ingredient category, according to the databases' scholarly records following the 2013 database updates. By individual food and food ingredients, the leading reported types of fraudulent foods were olive oil, milk, saffron, honey, coffee, tea, fish, clouding agents, and black pepper. Among the new reports examined, many of the most-represented products in the database were all among the top fraudulent products reported in the initial version of the database (such as milk, fish, turmeric, chili powder, and cooking oil); however, many products were not among the top products, such as shrimp, lemon juice, maple syrup, and clouding agents.[57]

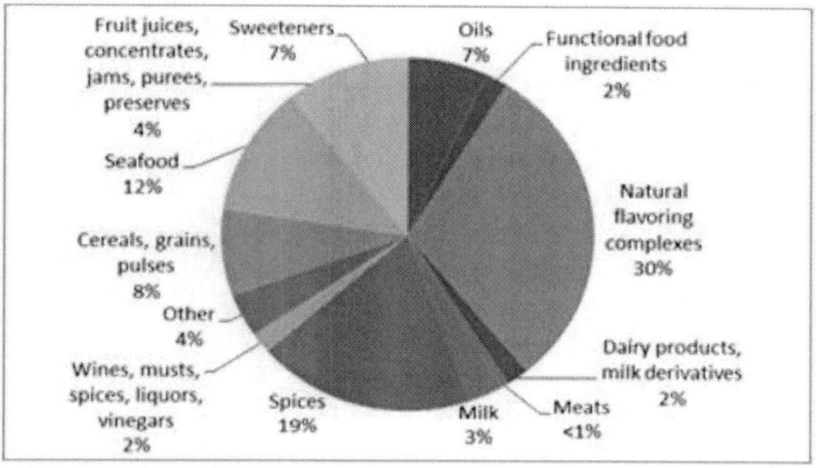

Source: USP Food Fraud Database (1980-2010), http://www.foodfraud.org/. As reported in J.C. Moore, J. Spink, and M. Lipp, "Development and Application of a Database of Food Ingredient Fraud and Economically Motivated Adulteration from 1980 to 2010," *Journal of Food Science*, Vol. 77, Nr. 4, 2012.

Notes: 1980-2010 data: records (251); ingredients (147); and references (85).

Figure 3. Leading Reported Types of Fraud, USP Media Records (1980-2010) By Food Ingredient Category, USP Food Fraud Database.

The 2013 updates to the USP database highlight a number of continuing and emerging issues. As reported by USP, among the leading reported types of fraud in the updated 2013 database are "watered-down and urea adulterated fluid milk in India, dilution of milk powder with fillers such as maltodextrin in South America and replacement of milk fat with vegetable oil in South America," as well as "olive oil replaced with other, less-expensive vegetable oils," and so-called "gutter oil" (waste oil repurposed as cooking oil) was documented in China.[58] The updated USP database also documents examples of the dilution or replacement of spices with less-expensive spices or fillers.

The updated USP database also highlighted increasing reports of seafood fraud, reflecting previous reports by Oceana and Consumers Union.[59] Of particular concern is mislabeled fish sold in the marketplace that may cause known public health risks. These include the fish escolar (*Lepidocybium flavobrunneum*), often labeled as white tuna or butterfish, which is banned in some countries due to concerns that its high content of waxy esters may cause some types of food poisoning including gempylotoxism.[60] Another is puffer fish (*Lagocephalus scleratus*), often mislabeled as monkfish to evade import and other restrictions, and which is known to cause tetrodotoxin poisonings.[61]

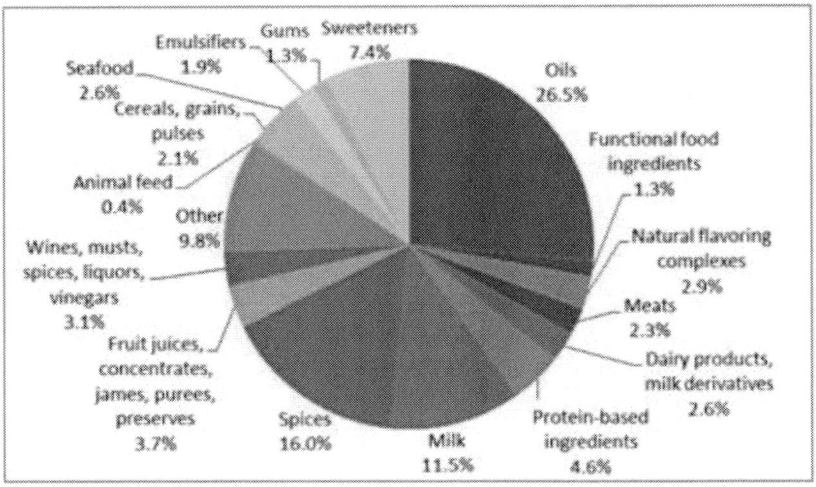

Source: Data from the USP Food Fraud Database (http://www.foodfraud.org/), 2013 update. As reported by A.G. Ebert, "The Food Chemicals Codex EMA Activities: The Food Fraud Database – What's Next?" Presentation at the USP Workshop of Economically Motivated Adulteration of Food Ingredients and Dietary Supplements, September 26–27, 2013. Data are for 2011-2012.

Notes: 1980-2012 data: records (1,648) and references (742). 2013 update included 792 additional records and 264 additional references, mostly from 2011-2012.

Figure 4. Leading Reported Types of Fraud, USP Scholarly Records (1980-2012) By Food Ingredient Category USP, Food Fraud Database.

The updated USP database also highlighted emerging concerns about food fraud involving "clouding agents." Specifically, the USP database documents cases involving the fraudulent addition of the plasticizer Di(2-ethylhexyl) phthalate (DEHP) and other related phthalates as a substitute for more expensive palm oil or other allowed food ingredients in fruit juices, jams, and other products. DEHP might also be used in food contact materials, such as seals and packaging. DEHP is associated with public health risks, including cancer and reproductive concerns. The scope of the fraud involving clouding agents covers 877 food products from 315 companies, with 206 products exported to a reported 22 countries.[62] Given the potential public health concerns, USP states that clouding agents might be considered the "2011 equivalent to the melamine scandal involving Chinese milk products from a few years ago."[63]

NCFPD EMA Incident Database

As of November 2013, about 300 "incidents" since 1980 are accessible in the National Center for Food Protection and Defense (NCFPD) EMA Incident Database.[64] Previous analyses of the incidents in the database were based on 137 incidents.[65] CRS analysis presented here is based on information from 302 incidents accessed in the EMA Incident Database, as well as other updated databases information from NCFPD.

As already noted, an incident is "a documented, isolated occurrence of EMA in a single food product or group of associated food products occurring within a defined time frame and with a distinct group of perpetrators."[66]

Figure 5 provides a breakdown of EMA incidents by major food ingredient category, according to the database. By individual food and food ingredients, the leading reported types of fraudulent foods were (1980 to date) fish and seafood (31%), oils and fats (11%), alcoholic beverages (8%), meat and meat products (7%), dairy products (6%), and about 5% each for grains and grain products, and honey and other natural sweeteners.

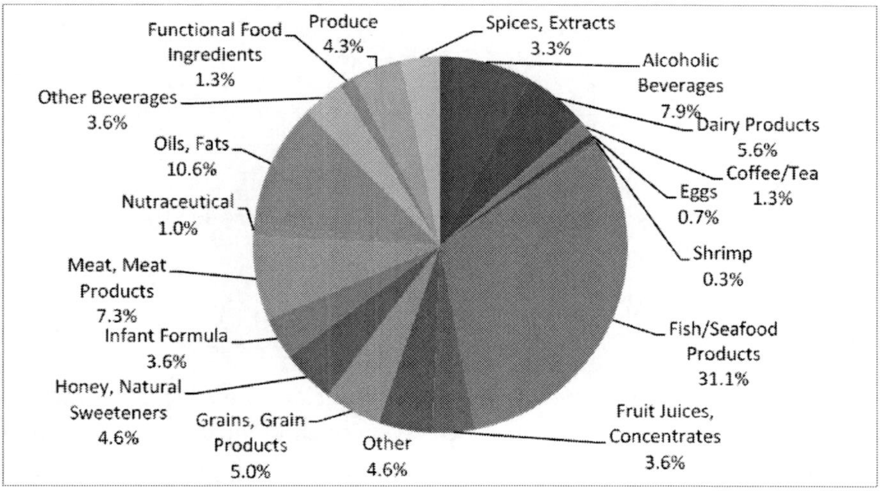

Source: Compiled by NCFPD CRS from records in the NCFPD EMA Incident Database (database accessed November 14, 2013) and based on 302 reported incidents. These incidents were also reported by A. Kircher, "Building Capabilities to Find and Mitigate." Presentation at the USP Workshop of Economically Motivated Adulteration of Food Ingredients and Dietary Supplements, September 26–27, 2013.

Figure 5. Leading EMA Incidents by Food Ingredient Category (1980 to date), NCFPD EMA Incident Database.

Figure 6 provides a breakdown of EMA incidents by type of adulteration, as reported in the database. Cases of EMA due to substitution or dilution accounted for 65% of the incidents, followed by 13% due to the presence of an unapproved additive. Other incidents are attributable to counterfeit goods (9%), misbranding (7%), transshipment or masking origin of product (5%), and the intentional distribution of a potentially hazardous substance (less than 1%), among other miscellaneous or unknown types of adulteration (less than 1%).

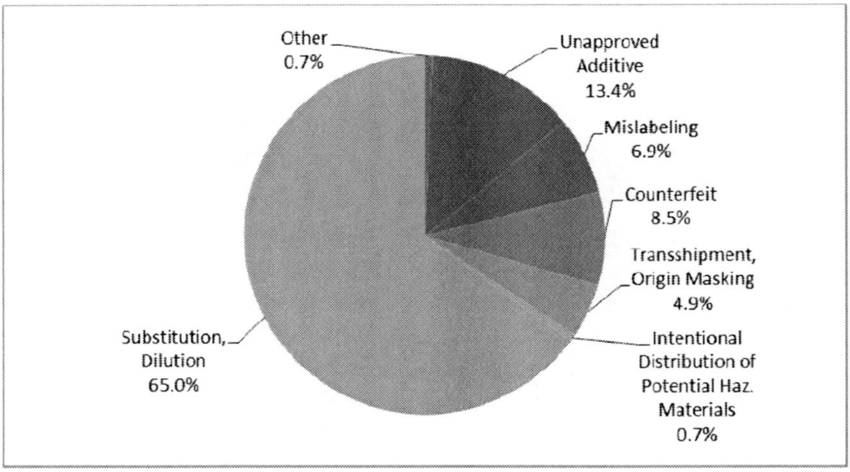

Source: Compiled by NCFPD EMA Incident Database (January 6, 2014), based on 306 reported incidents.

Figure 6. Leading EMA Incidents by Type of Adulteration (1980 to date), NCFPD EMA Incident Database.

Figure 7 provides a breakdown of EMA incidents by location where the product is identified as being produced, as reported in the database. Among its data, the EMA Incident Database documents where the fraudulent product was originally produced based on an examination of the available information. Although the database is not restricted to conditions in the United States and is broadly reflective of fraud globally, the incidents captured in the database are based on English-language sources and may not be representative of conditions across all global commercial markets.

In addition, information in the database by location produced may mask substantial differences among countries and regions (both globally and within a particular country) in terms of their ability and dedication to detect fraud. No two countries or regions scrutinize food fraud in the same way. For example, as noted by researchers at NCFPD,[67] the relatively greater number of reported fish and seafood cases is, in part, attributable to enhanced surveillance and detection by Florida's Department of Health. This further reinforces the idea that a possible limitation of either database is that the available information may at times be more reflective of where research is being conducted or where resources have been dedicated to detect fraud.

Despite these caveats, the EMA Incidents Database indicates that nearly 30% of the EMA incidents documented involved products that were produced in the United States. Most of these were incidents involving mislabeled fish. China and India accounted for another 14% and 13%, respectively, of the documented EMA incidents. Other countries in Asia accounted for another 5% of the incidents. Combined, the European Union countries accounted for about 15% of the documented incidents, along with another 3% attributable to other non-EU European countries.

About 7% of the incidents were reported to have been produced in the Middle East and Africa (Figure 7).

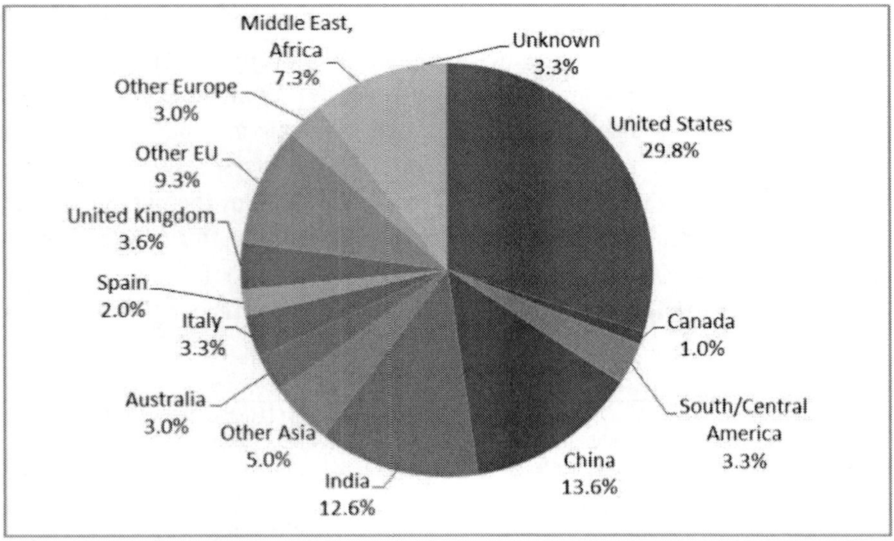

Source: Compiled by CRS from records in the NCFPD EMA Incident Database (database accessed November 14, 2013) and based on 302 reported incidents.

Figure 7. Leading EMA Incidents by Location Produced (1980 to date), NCFPD EMA Incident Database.

Differences in Product Ranking among Databases

Reported rankings of foods and food ingredients associated with food fraud based on the USP Food Fraud Database and the NCFPD EMA Incident Database often differ. As illustrated above, how these fraudulent foods and food ingredients are ranked will differ depending on which database is referenced (i.e., whether USP or NCFPD database) and which type of record is referenced (i.e., whether based on "scholarly" or "media" or EMA "incidents"), and time period.

Rankings also differ within a database. For example, as reported in the 2013 food fraud report by the European Parliament, the "top ten products that are most at risk of food fraud" using data in the initial USP database (1980-2010) are identified (in descending order) as olive oil; fish; organic foods; milk; grains; honey and maple syrup; coffee and tea; spices (such as saffron and chili powder); wine; and certain fruit juices.[68] Comparisons, shown in Table 3, are intended to illustrate why different reports provide differing rankings of foods associated with fraud. The prominence of "organic foods" in the European study may be explained by the decision to distinguish organically certified products separately from all reported foods categorized by food ingredient category (which include both organic and conventionally produced foods and food ingredients). The overall prominence of "fish" in the European study may also be explained in that products may have been ranked across all records and not separated according to "scholarly" versus "media" records. Although these compilations were derived from the same database, this example illustrates why some reported compilations differ from others.

Table 3. Leading Reported Types of Food Fraud, Differing Compilations Various Years and Sources

USP Food Fraud Database: Leading Ingredient Categories, Accounting for >90% of Records (Scholarly and Media Combined)		NCFPD EMA Incidents Database: Leading Ingredient Categories, Accounting for >90% of Records (Media)		EU Report: "Top ten" products at risk of food fraud
1980-2010	1980-2012	1980-2010	1980-2012	1980-2010
(a)	(b)	(c)	(d)	(e)
Oils	Oils	Other miscellaneous	Fish, seafood	Olive oil
Spices	Spices	Fish and seafood	Oils and fats	Fish
Milk	Milk	Dairy products	Alcoholic beverages	Organic foods
Fruit juice, concentrate	Other miscellaneous	Fruit juices	Meat, meat products	Milk

USP Food Fraud Database: Leading Ingredient Categories, Accounting for >90% of Records (Scholarly and Media Combined)		NCFPD EMA Incidents Database: Leading Ingredient Categories, Accounting for >90% of Records (Media)		EU Report: "Top ten" products at risk of food fraud
1980-2010	1980-2012	1980-2010	1980-2012	1980-2010
(a)	(b)	(c)	(d)	(e)
Natural flavorings	Sweeteners	Oils and fats	Dairy products	Grains
Sweeteners	Fruit juice, concentrate	Grain products	Grains, grain products	Honey, maple syrup
Other miscellaneous	Natural flavorings	Honey	Honey, sweeteners	Coffee, tea
Cereals, grains, pulses	Seafood (incl. fish)	Spices and extracts	Produce	Spices
Seafood	Cereals, grains, pulses	Wine	Fruit juice, concentrate	Wine
Dairy products	Wines, musts, spirits	Infant formula	Spices, extracts	Fruit juices
Wines, spirits, vinegars	Dairy products	Plant-based proteins	Other beverages (nonalcoholic, not juice)	
	Protein-based ingred.			
	Functional food ingred.			
	Preservatives			
	Gums			
	Meats			

Sources:

(a), (b) Compiled by Jeff Moore, USP, December 22, 2013.

(c) NCFPD presentation, USP's Food Ingredients Intentional Adulterants Expert Panel, Meeting #3, October 23, 2012. Provided by Jeff Moore, USP, December 2013.

(d) Compiled by Karen Everstine, NCFPD, January 2014.

(e) As reported in European Parliament, Public Health and Food Safety draft report "on the food crisis, fraud in the food chain and the control thereof," (2013/2091(INI)), citing information obtained from the USP database.

Notes: Data from the USP Food Fraud Database and the NCFPD EMA Incident Database. a. Clouding agents are used as food processing aids (21 CFR Part 170, Food Additives).

FEDERAL ACTIVITIES INVOLVING FOOD FRAUD

In the United States, no single federal agency and no single U.S. law or statute directly addresses food fraud or "economically motivated adulteration" of food and food ingredients. Instead, food fraud and intentional adulteration of food is broadly addressed through food safety authorities and border protection and import authorities and activities. Accordingly, FDA and USDA are the principle federal agencies that are working to protect the food supply from food safety risks, including both unintentionally and intentionally introduced contamination, in conjunction with enforcement of FDA and USDA laws by the U.S. Department of Homeland Security (DHS) as part of its border inspections.

FDA and USDA are the leading food safety regulatory authorities. FDA, an agency of the Department of Health and Human Services (HHS), is responsible for ensuring the safety of all domestic and imported food products (except for most meats and poultry). USDA's Food Safety and Inspection Service (FSIS) regulates most meat and poultry and some egg products. However, the boundaries between the two agencies' jurisdictions are complex and often at odds with commonplace distinctions among food groups.[69] The laws that grant these and other federal agencies authority over foods and food ingredients also provide these agencies with the authority to govern the labeling of these products, which might also act as a further deterrent to food fraud in some cases.[70]

Other federal agencies also provide product quality standards and grading, including FDA, USDA's Agricultural Marketing Service (AMS), and the National Marine Fisheries Service (NMFS), which is part of the U.S. Department of Commerce's National Oceanic and Atmospheric Administration (NOAA). State and local food safety authorities assist the federal agencies with inspection, outbreak response, and other food safety functions, among other functions. Import security measures by FDA and USDA are conducted in conjunction with border inspections by the Customs and Border Protection (CBP), which is part of DHS.

Other agencies also have played a role in food fraud prevention. The U.S. Department of Justice (DOJ) has actively pursued a number of food fraud cases in the U.S. courts involving a range of products and resulting in criminal convictions in some cases.[71] In the high-profile case involving the Peanut Corporation of America, company executives were indicted in early 2013 with federal criminal charges that included fraud, conspiracy, and the introduction of adulterated food into interstate commerce with the intent to defraud or mislead.[72] USDA's National Organic Program (NOP) has taken numerous enforcement actions against companies involving false labeling claims on organic foods, and a number of advocacy groups have filed lawsuits involving concerns about the validity of organic claims. NOAA also has conducted federal investigations and taken enforcement actions involving seafood labeling and misbranding.[73]

The text box below provides a comparison of FDA and USDA responsibilities for food safety and regulations, and responsibilities of other federal agencies.

Comparison of Selected Agency Responsibilities for Food Safety and Regu

Agency	Responsibility
Food and Drug Administration (FDA)	• Food (but not meat)
	• Dietary supplements
	• Bottled water
	• Seafood
	• Wild game (" exotic " meat)
	• Eggs in the shell
U.S. Department of Agriculture (USDA)	• Grading of raw fruit and vegetables
	• Meat and Poultry • Eggs, processing and grading • Certifying organic production
National Oceanic and Atmospheric Administration	• Grading of fish and seafood
Environmental Protection Agency (EPA)	• Drinking water
	• Pesticide residues
Customs and Border Protection (CBP)	• Front-line enforcement and refe
Department of Justice (DOJ)	• Law enforcement
Federal Trade Commission (FTC)	• Advertising
Alcohol and Tobacco Tax and Trade Bureau (TTB)	• Alcohol

Source: CRS, as adapted from N. D. Fortin, *Introduction to Food Regulation in the United States,* Part 1 (Introductory Chapters), May 2008.

The federal government also has helped fund some of the ongoing research. These efforts were developed with the support of funding from DHS, FDA, and USDA.[74] In addition, FDA is developing a number of tools and guidance materials for the food industry to address intentional adulteration as part of its food defense strategy.[75]

The U.S. Government Accountability Office (GAO) has conducted a series of investigations into food fraud cases, including fruit juice[76] and seafood.[77] GAO's most recent report addressed a wider range of fraudulent foods and issued more broad-based recommendations.[78] GAO's recommendations involving food fraud addressed the need for improved cross-agency communication and coordination and enhanced information sharing and transparency. Other recommendations called for the creation of an information clearinghouse and the need for increased oversight and inspections, and increased risk-based testing, in some cases.

A 2013 report by the U.S. International Trade Commission (USITC) cited research indicating that current standards for extra virgin olive oil are widely unenforced and result in a wide range of olive oil qualities to be labeled as "extra virgin" oil. The study further concluded that this may lead to adulterated and mislabeled product, which could further weaken the competitiveness of U.S.-produced olive oil in the domestic market.[79]

In Europe, in the wake of the horsemeat scandal, some countries have debated the need for tougher laws to protect consumers against food fraud. The European Parliament released its draft report regarding food fraud in October 2013, calling for increased enforcement and oversight regarding fraud prevention through the food supply chain.[80]

HHS, Food and Drug Administration

Laws governing FDA's authority over both unintentional and intentional adulteration of both domestic and imported food—namely, the Federal Food, Drug, and Cosmetic Act—provide the agency with some of the necessary tools to pursue activities that ensure against fraud of foods and food ingredients under its jurisdiction.

Food Safety Authorities

FDA has primary responsibility for the safety of most—about 80%-90%—of all U.S. domestic and imported foods.[81] The FDA is responsible for ensuring that all domestic and imported food products—except for most meats and poultry—are safe, nutritious, wholesome, and accurately labeled. Examples of FDA-regulated foods are produce, dairy products, and processed foods.

FDA also has oversight of all seafood and shellfish products, and most fish products.[82] FDA has jurisdiction over meats from animals or birds that are not under the regulatory jurisdiction of FSIS. FDA shares some responsibility for the safety of eggs with FSIS. FDA has jurisdiction over establishments that sell or serve eggs or use them as an ingredient in their products.

As described in a Memorandum of Understanding between FDA and FSIS:[83]

> FDA is responsible for implementing and enforcing the Federal Food, Drug, and Cosmetic Act (21 U.S.C. 301, *et seq.*), the Public Health Service Act (42 U.S.C. 201, *et seq.*), the Fair Packaging and Labeling Act (15 U.S.C. 1451 *et seq.*), and parts of the Egg Products Inspection Act. In carrying out its responsibilities under these acts, FDA conducts inspections of establishments that manufacture, process, pack, or hold foods, with the exception of certain establishments that are regulated exclusively by FSIS. FDA also inspects vehicles and other conveyances, such as boats, trains, and airplanes, in which foods are transported or held in interstate commerce.

"Prohibited" Acts

FDA's food safety authorities rest primarily with the Federal Food, Drug, and Cosmetic Act (FFDCA), as amended, requiring that foods be safe, wholesome, and accurately labeled. "Prohibited" acts are listed in FFDCA Section 301 (21 U.S.C. §331). Along with other specified prohibited acts, the law provides that introducing adulterated or misbranded food into commerce; adulterating or misbranding food that is in commerce; or the receipt and delivery of adulterated or misbranded food in commerce is prohibited. As such, under FFDCA, two of the basic statutory components are "adulteration" and "misbranding."

FDA-regulated foods may be deemed adulterated or misbranded for a variety of statutorily prescribed reasons. For example, food may be deemed adulterated if it contains an added poisonous or deleterious substance or an unsafe food additive or if the food was prepared, packed, or held under unsanitary conditions whereby it may have become contaminated or may have been rendered injurious to health. Persons who violate FFDCA by, for example, introducing an adulterated or misbranded product into interstate commerce are subject to criminal and civil penalties.

According to FFDCA Section 402(a) (21 U.S.C. §342[a]), a food shall be deemed to be "adulterated" (and therefore prohibited) if it bears or contains or has any added "poisonous or deleterious substance" which may be "injurious to health" or is considered unsafe under the law, unless specifically exempted. Adulteration also covers food that consists of a "filthy, putrid, or decomposed substance," or if it is otherwise unfit for food or "if it has been prepared, packed, or held under insanitary conditions," or if it otherwise poses a risk to consumer health.

Perhaps more specific to food fraud, although generally less noted,[84] FFDCA Section 402(b) (21 U.S.C. §342[b]), are foods deemed to be adulterated given the "absence, substitution, or addition of constituents" including:[85]

> if any valuable constituent has been in whole or in part omitted or abstracted therefrom; or (2) if any substance has been substituted wholly or in part therefor; or (3) if damage or inferiority has been concealed in any manner; or (4) if any substance has been added thereto or mixed or packed therewith so as to increase its bulk or weight, or reduce its quality or strength, or make it appear better or of greater value than it is.

Accordingly, a food may be deemed adulterated if it is missing a "valuable constituent" or if substances have been substituted wholly or in part or where damage or inferiority has been concealed in any matter, including cases where substances are added to "increase its bulk or weight" or make it appear to be higher quality.

Under FFDCA, introducing misbranded food into commerce, misbranding food that is in commerce, or the receipt and delivery of misbranded food in commerce is also prohibited. FFDCA Section 403 (21 U.S.C. §343) defines a number of conditions under which a food would be deemed to be misbranded. A food is deemed misbranded if it has a false or misleading label, is offered for sale under another name, is an imitation of another food, or is in a "misleading container" that is "made, formed, or filled to be misleading," among other types of misrepresentative packing or labeling. Similar to the definition of adulteration, numerous specific types of misbranding are also defined. These include, among others, failure to disclose specific additives or allergens in the food, and failure to provide required nutritional information. (In addition to FFDCA, FDA has regulatory authority under the Fair Packaging and Labeling Act [FPLA, 15 U.S.C. §1451 *et seq.*]. FPLA establishes requirements for package labels of all consumer goods, including most foods.)

Violating the misbranding or adulteration provisions of FFDCA may invoke criminal penalties and/or product seizures. Under FFDCA Section 303 (21 U.S.C. §333), in general, any person who commits a prohibited act under FFDCA may be subject to civil or criminal penalties, including imprisonment, fines, or both. Criminal penalties provided for in FFDCA are adjusted by 18 U.S.C. Sections 3559 and 3571. Certain exceptions may be made, including for the misbranding of foods. FFDCA Section 304 (21 U.S.C. §334) also provides that FDA may order the administrative detention of foods, if a food is suspected to be adulterated or misbranded based on an inspection, examination, or investigation.

Although fraudulent activities may be illegal and enforceable under the FFDCA, FDA might not take enforcement in some cases. Practically speaking, it may not be possible for FDA and DOJ to prosecute every instance of food fraud given each agency's myriad other responsibilities and limited personnel and resources. Also, oftentimes inadequate evidence exists to effectively enforce against all alleged or suspected cases of fraud.

Other Agency Authorities and Activities

FFDCA Section 415 (21 U.S.C. §350d) requires FDA to oversee the registration of both domestic and foreign food facilities, which was part of the Public Health Security and Bioterrorism Preparedness and Response Act of 2002 ("Bioterrorism Act").[86] Under the act, facilities that manufacture, process, pack, or hold food for human or animal consumption in the United States must be registered with FDA.[87] Domestic facilities must register whether or not food from the facility enters interstate commerce. Foreign facilities must register unless food from that facility undergoes further processing (including packaging) by another foreign facility before the food is exported to the United States. FDA also must be given advance notice on shipments of imported food. These registration and prior notification requirements might act as a further deterrent to food fraud in some cases.

Table 4 shows available data on the total number of registered domestic and foreign food facilities. The number of registered food facilities—both domestic and foreign—more than doubled from 2004 to 2012, the most recent available data.[88] Of the total number of registered facilities with FDA in 2012 (nearly 450,000), a reported 171,552 were domestic facilities and 278,307 were foreign facilities (Table 4).

Finally, FFDCA also broadly provides for FDA to establish voluntary food quality and grading standards, and product standards of identity (FFDCA §401 [21 U.S.C. §341]). Although not regulatory in nature, and not intended to address potential food safety or food fraud concerns, food quality and grading standards provide a product benchmark for certain foods or food ingredients. Specifically, FFDCA directs FDA to establish definitions and standards for food to "promote honesty and fair dealing" for the benefit of consumers.[89] Under the statute, FDA is authorized to establish regulations "for any food ..., a reasonable definition and standard of identity, a reasonable standard of quality, and reasonable standards of fill" of the container for any food. FDA has established roughly 300 identity standards in 20 categories of food, consisting of a range of processed foods and meat, dairy, and seafood products, as well as preserved and processed fruit and vegetable products and juices. Standards of identity cover mostly processed and value-added foods for a wide range of FDA-regulated food products.

Table 4. Registered Food Facilities, FY2004-FY2012

	FY04	FY05	FY06	FY07	FY08	FY09	FY10	FY11	FY12
All Registered Food Facilities	214,253	253,006	288,092	323,590	356,287	391,281	418,593	438,305	449,859
Domestic	92,719	104,555	115,902	129,345	141,703	154,883	166,160	167,033	171,552
Foreign	121,534	148,451	172,190	194,245	214,584	236,398	252,433	271,272	278,307

Source: Compiled by CRS from data on registered domestic and foreign facilities under FFDCA §415 (21 U.S.C. §350d); FDA's annual reporting requirements of these data are at FFDCA §1003 (21 U.S.C. §393).

Notes: Number of registrants as of November 18, 2013. Available FY2012 data are from FDA, "Registration Statistics," http://www.fda.gov/food/guidanceregulation /foodfacilityregistration/ucm236512.htm. FY2004-2011 data are FDA Office of Legislation (September 22, 2010, and May 7, 2012, communication); FDA, "Annual Report on Food Facilities, Food Imports, and FDA Foreign Offices" for 2011 and 2012, http://www.fda.gov/Food/ GuidanceRegulation/FSMA/ucm 250569.htm; and HHS OIG, *FDA Inspections of Domestic Food Facilities* (OEI-02-08- 00080).

Potential Role of the Food Safety Modernization Act

FDA has taken a series of internal actions to address intentional adulteration, including EMA.[90] In April 2009, FDA conducted a public meeting to raise awareness and solicit public input regarding economically motivated adulteration of FDA-regulated products. FDA also established, in September 2011, an internal workgroup comprised of staff from all FDA product centers (covering both food and non-food products regulated by FDA), FDA's Office of Regulatory Affairs (ORA),[91] and the Office of the Commissioner. The workgroup includes risk managers, economists, regulatory counsel, policy analysts, and scientists, and uses a multidisciplinary collaborative approach to capitalize on commonalities among FDA's product centers.[92] Part of the workgroup's mission is to address recommendations regarding EMA from the U.S. Government Accountability Office (GAO) that called for improved cross-agency communication and enhanced information sharing and transparency with stakeholders, including the regulated community.[93] FDA reportedly is also considering another GAO recommendation that would establish an information clearinghouse so stakeholders could share information about products that may be susceptible to economic adulteration.[94]

The 111th Congress amended FFDCA by passing a comprehensive food safety law, the Food Safety Modernization Act (FSMA, P.L. 111-353).[95] FSMA aims to prevent both intentional and unintentional introduced contamination of foods through a variety of strategies to prevent food contamination and through enhanced regulatory authorities. FDA has not yet implemented many of the law's major provisions.[96] FSMA provides for increased food risk protection by creating mechanisms whereby food companies are required to identify and implement preventive controls to ensure adulterated products are not sold and then share their plan with the FDA to ensure compliance with good manufacturing practices. Having such controls in place would also allow companies to consider their responsibility regarding potentially adulterated foods involving the "absence, substitution, or addition of constituents." Some speculate that provisions under FSMA will likely result in increased detection and prevention of adulterated foods by the food companies, both under FFDCA Section 402(a) (21 U.S.C. §342[a]) and FFDCA Section 402(b) (21 U.S.C. §342[b]).[97] Since FDA could not possibly enforce every instance of food adulteration, such industry controls will be instrumental in helping to combat future adulteration before it ever reaches the marketplace.

FSMA does not directly address EMA; however, some of its provisions may have application to EMA even though some of these provisions may have originated as part of an overall food defense strategy. These FSMA provisions include:

- **FSMA Section 103 (Hazard Analysis and Risk-Based Preventive Controls).**[98] FSMA requires preventive controls for human food by domestic and foreign firms that manufacture, process, pack, or hold human food. Among the requirements are written plans that identify hazards, specify the steps that will be put in place to minimize or prevent those hazards, identify monitoring procedures, and record monitoring results and specify what actions will be taken to correct problems that arise. These regulations would address, among other things, "hazards that occur naturally, may be unintentionally introduced, or may be intentionally introduced, including by acts of terrorism."

As part of its proposed rule published in January 2013, FDA states that intentional hazards, such as EMA, might require separate agency action. As part of the proposed rulemaking, FDA also has requested public comment on if and how it should address EMA:[99]

> FDA tentatively concludes that intentional hazards, which are not addressed in traditional HACCP[100] or other food safety systems, likely will require different kinds of controls and would be best addressed in a separate rulemaking ... We request comment on whether to include potential hazards that may be intentionally introduced for economic reasons ... [and] on when an economically motivated adulterant can be considered reasonably likely to occur.

- **FSMA Section 106 (Protection Against Intentional Adulteration).** FSMA requires FDA to issue regulations to protect against the intentional adulteration of food, such as "specifying appropriate science-based mitigation strategies or measures to prepare and protect the food supply chain from intentional adulteration at specific vulnerable points."[101] The proposal for this rulemaking was published in December 2013, and is the first time FDA has proposed a regulatory approach for preventing intentional adulteration of food.[102]
- **FSMA Section 108 (National Agriculture and Food Defense Strategy).** FSMA requires that the Secretaries of Health and Human Services and Agriculture develop a National Agriculture and Food Defense Strategy, implementation plan, and research agenda. This strategy and the accompanying documents have not yet been published.[103]
- **FSMA Title 3 (Improving the Safety of Imported Food).** To the extent that food fraud may be attributable to foods imported into the United States, FSMA provides for a series of requirements to ensure the safety of imported foods. Tightened import requirements may act to limit future fraud, or lend greater credibility to foods and food ingredients that are imported under these requirements. FSMA also recognizes "third party" audits or certifications, and several such entities have already started to address food fraud both in term of identifying terms or assessing implementation actions such as vulnerability assessments.[104] FDA's proposed rules were published in July 2013.[105]
- **FSMA Section 402 (Employee Protections).** FSMA expands protections for employees (whistleblowers) who provide information

relating to FFDCA violations, such as testifying, assisting, or participating in a proceeding on such a violation, refusing to participate in an activity that may violate FFDCA.

- **FSMA Section 201(b) (Annual Report Regarding Food).** FSMA requires FDA to submit an annual to report to Congress, covering efforts to coordinate and cooperate with other federal agencies with responsibilities for food inspections, and report information regarding facility inspections and facility registrations, among other things.[106]

FDA has been working with other federal agencies to coordinate and cooperate on food safety activities. According to FDA's most recent report on these types of activities, for example, FDA and the Department of Defense have "information-sharing networks and processes on facility audits and inspections, recalls, import alerts, laboratory findings and methods, and other food protection procedures."[107] FDA officials have indicated that obtaining additional data and information would likely require enhanced public and private partnerships.[108]

Import Authorities

Import security measures by FDA are conducted in conjunction with border inspections by CBP.[109] As part of this responsibility, importers of foreign food are responsible for verifying that the products obtained from foreign processors are in compliance with U.S. laws. FDA's surveillance of imported foods consists of reviews of prior notice data, reviews of customs entry forms, physical or sensory analysis, sample collections for laboratory analysis, and detention without physical examination.

As required by the Bioterrorism Act, FDA must have received a notice for articles of food being imported or offered for import into the United States, prior to importation (FFDCA §801(m) [21 U.S.C. §381(m)]). Prior notice is required to enable the food to be inspected at U.S. ports of entry, and FDA must refuse admission to food imported or offered for import if the notice was not submitted or if the notice was deficient. Additionally, FDA may hold food at the port of entry if it is imported or offered for import by a person who was debarred under FFDCA or if it was imported or offered for import from a foreign facility that has not registered with the FDA. FDA screens the electronic shipping records of all imported food products before they enter the United States. From these records, the agency selects products for physical examination and/or testing to determine whether they contain adulterants.

Under FFDCA Section 801 (21 U.S.C. §381), FDA has the authority to refuse entry of any food import if it appears to be adulterated, misbranded, or in violation of U.S. law. In such cases, FFDCA generally provides that such a food article must be refused admission into the United States, with few exceptions, "if it appears from the examination of such samples or otherwise" that it has been "manufactured, processed, or packed under insanitary conditions," or it is "forbidden or restricted in sale in the country in which it was produced or from which it was exported," or it is "prohibited from introduction or delivery for introduction into interstate commerce under section 301(ll)" (FFDCA §801(a)).

FDA actions on suspect imported products may be implemented through FDA's "Import Alerts." An alert can be issued for an import from a manufacturer, shipper, grower, geographical area, or country. An active import alert provides a signal to border inspectors to look more closely at a particular product, or a range of products from a particular producer, shipper, or importer. If the problem or condition exists on a wide scale, FDA may be instructed to detain all products of a certain kind coming from a country or a region of a country. Products that may be subject to refusal based on existing evidence (such as a history of violations) can be detained at the border and refused admission into U.S. commerce unless the importer is able to demonstrate that the products are in compliance.[110] Import alerts may allow U.S. authorities (typically the U.S. Treasury has delegated its authority via DHS to the Customs and Border Protection) to detain, without physically examining, products that either have or potentially could violate FFDCA.[111] Detention without physical examination (DWPE), formerly known as "automatic detention," was developed to address recurrent violations.[112]

In FY2011, FDA reports that it physically examined (conducted field exams or analyzed samples of U.S. food imports under its jurisdiction) 243,400 food and feed import lines.[113] This was out of a total of 10.4 million food import lines for FY2011.[114] Hence, FDA physically examined 2.3% of all shipments. As noted by FDA, this physical examination was in addition to FDA's electronic screening of all import lines against a variety of risk criteria. As part of FDA's screening process, it implemented its Predictive Risk-based Evaluation for Dynamic Import Compliance Targeting (PREDICT) information technology system that helps target high-risk products before they enter the United States. Widely cited previous estimates of FDA food safety inspections of foreign facilities indicate that the rate of such inspections was even lower: GAO reported that, in 2000, FDA inspections covered only about

1% of the food imported under its jurisdiction.[115] Changes to FDA's import regime under FSMA are expected to further address some of these concerns.

USDA, Food Safety and Inspection Service

Similar to laws governing FDA that may be broadly construed to provide the agency's authority over both unintentional and intentional adulteration of the food supply, the principal food safety laws governing USDA—namely, the Federal Meat Inspection Act and the Poultry Products Inspection Act—might likewise allow USDA to pursue activities to ensure against fraud of meat and poultry products under USDA's jurisdiction.

Food Safety Authorities

Excluding FDA-regulated foods, FSIS is responsible for the safety of the remaining roughly 10%- 20% of foods covered by the U.S. food safety system, including meat and poultry and some egg products.[116] As described in a Memorandum of Understanding between FDA and FSIS, FSIS's jurisdiction is as follows:[117]

> FSIS is responsible for implementing and enforcing the Federal Meat Inspection Act (21 U.S.C. 601, *et seq.*), the Poultry Products Inspection Act (21 U.S.C. 451, *et seq.*), and parts of the Egg Products Inspection Act (21 U.S.C. 1031, *et seq.*). In carrying out its responsibilities under these acts, FSIS places inspectors in meat and poultry slaughterhouses and in meat, poultry, and egg processing plants. FSIS also conducts inspections of warehouses, transporters, retail stores, restaurants, and other places where meat, poultry, and egg products are handled and stored. In addition, FSIS conducts voluntary inspections under the Agriculture Marketing Act (7 U.S.C. 1621, *et seq.*).

The Federal Meat Inspection Act (FMIA) of 1906, as amended, requires USDA to inspect all cattle, sheep, swine, goats, horses, mules, and other equines slaughtered and processed for human consumption. The Poultry Products Inspection Act (PPIA) of 1957, as amended, gives USDA the authority to inspect poultry meat. The PPIA mandates USDA inspection of any domesticated birds (chickens, turkeys, ducks, geese, guineas, ratites [ostrich, emu, and rhea], and squab [pigeons up to one month old]) intended for use as human food. The Egg Products Inspection Act, as amended, provides USDA authority to inspect liquid, frozen, and dried egg products. Each of these laws also contain provisions governing USDA's authority to label food products under its jurisdiction.[118]

Under the authority of the Agricultural Marketing Act of 1946 as amended, USDA's FSIS may provide voluntary inspection for buffalo, antelope, reindeer, elk, migratory waterfowl, game birds, and rabbits. This type of inspection is performed by FSIS on a fee-for-service basis. However, these meat and poultry species are still within the purview of FDA under FFDCA, whether or not inspected under the voluntary FSIS program. FDA has jurisdiction over meat products from such species in interstate commerce, even if they bear the USDA inspection mark. FDA also has jurisdiction over shell eggs. In addition, the 2008 farm bill requires that FSIS inspect and grade farmed catfish products.[119]

Meat and poultry animals and products undergo continuous (i.e., 100%) inspection, which may in turn act as a deterrent to fraud in some cases. FSIS inspects all meat and poultry animals to look for signs of disease, contamination, and other abnormal conditions, both before and after slaughter ("antemortem" and "postmortem," respectively), on a continuous basis—meaning that no animal may be slaughtered and dressed unless an inspector has examined it. One or more federal inspectors are on the line during all hours the plant is operating. Processing plants visited once every day by an FSIS inspector are considered to be under continuous inspection in keeping with the laws. Inspectors monitor operations, check sanitary conditions, examine ingredient levels and packaging, review records, verify food safety plans,[120] and conduct statistical sampling and testing of products for pathogens and residues during their inspections.[121]

FSIS is responsible for certifying that foreign meat and poultry plants are operating under an inspection system equivalent to the U.S. system before they can export their product to the United States.[122] Meat and poultry imports are 100% visually inspected (process-based, documentation, labeling), although physical inspections of imports may be more random. FSIS conducts evaluations of foreign meat safety programs and visits establishments to determine whether they are providing a level of safety equivalent to that of U.S. safeguards. No foreign plant can ship meat or poultry to the United States unless its country has received such an FSIS determination.

About 1,100 of the establishments under the jurisdiction of FSIS either slaughter, or slaughter and process livestock, or poultry.[123] More than 4,000 facilities only process meat and poultry, and about 80 process egg products. In addition to inspecting domestic meat, poultry, and egg establishments, FSIS also performs re-inspections of imported meat, poultry, and egg products at about 140 import re-inspection facilities.

"Prohibited" Acts

FMIA and PPIA also authorize USDA to regulate labeling and packaging of meat, poultry, or processed parts to prevent false or misleading marks, labels, or containers (FMIA, 21 U.S.C. §607 and PPIA, 21 U.S.C. §457).[124] Similar to FFDCA, both FMIA and PPIA also disallow certain "prohibited" acts that involve products for use as human food that are "adulterated or misbranded at the time of such sale, transportation, offer for sale or transportation, or receipt for transportation" or products "required to be inspected ... unless they have been so inspected and passed" (FMIA, 21 U.S.C. §10; PPIA, 21 U.S.C. §458).

FSIS is responsible for developing the labeling policy to determine that meat or poultry products are wholesome, not adulterated, and properly marked, labeled, and packaged. FMIA and PPIA both define "misbranded" foods as bearing a false or misleading label, or foods that are "offered for sale under the name of another food," or are "an imitation of another food, unless its label bears, in type of uniform size and prominence," or if "its container is so made, formed, or filled as to be misleading, or is otherwise misrepresented" (21 U.S.C. §§ 453 and 601). An "adulterated" food bears or contains any "poisonous or deleterious substance which may render it injurious to health" or otherwise poses a risk to consumer health.

Perhaps more specific to food fraud, both FMIA and PPIA also define adulteration to include the following:[125]

> if any valuable constituent has been in whole or in part omitted or abstracted therefrom; or if any substance has been substituted, wholly or in part therefor; or if damage or inferiority has been concealed in any manner; or if any substance has been added thereto or mixed or packed therewith so as to increase its bulk or weight, or reduce its quality or strength, or make it appear better or of greater value than it is.

Accordingly, similar to FFDCA, a food may be deemed adulterated if a "valuable constituent" is missing or if substances have been substituted wholly or in part or where damage or inferiority has been concealed in any matter, or if substances have been added to "increase its bulk or weight" (such as packing water or ice) or to raise the perceived quality of the product.

Moreover, FMIA and PPIA and federal meat and poultry inspection regulations (9 C.F.R. §§301.2 and 381.1, respectively) define "meat," "meat food product," "livestock," "poultry," and "poultry product" and do not include species of livestock or poultry other than those specifically listed. FSIS verifies the labeling, conducts species testing, and monitors the movement in commerce of all meat and poultry food products produced under federal inspection. This includes all meat and poultry products as well as other meat products (such as horsemeat and meat from various other exotic animals) produced in or imported into the United States. FSIS claims that its stringent inspection process, testing capabilities, and labeling requirements effectively prevent cases such as those that occurred in the EU earlier in 2013, where horsemeat was labeled and sold as the meat of another species. FSIS claims it "conducts species tests on meat and poultry products that are produced domestically and that are imported to the United States" which are "capable of detecting beef, sheep, swine, poultry, deer and horse."[126]

Other Agency Authorities and Activities

Finally, FMIA and PPIA also broadly provide for FSIS to establish voluntary standards for meat and poultry products (FMIA, 21 U.S.C. §607[c]; and PPIA, 21 U.S.C. §457[b]). Although not regulatory in nature, and not intended to address potential food safety or food fraud concerns, food quality and grading standards provide a product benchmark for certain foods or food ingredients. Specifically, both FMIA and PPIA direct USDA to establish "definitions and standards of identity or composition" for meat and poultry products.[127] Standards of identity cover a wide range of raw, cooked, cured, and processed meat and poultry products and ingredients for USDA-regulated food products.

Import Authorities

Similar to that for FDA, various import security measures by USDA are conducted via various enforcement and border inspection activities by CBP. As part of this responsibility, importers of foreign food are responsible for verifying that the products obtained from foreign processors are in compliance with U.S. laws.

DHS, Customs and Border Protection

CBP enforces FDA and USDA regulations at ports of entry. Import security measures, in conjunction with existing CBP border inspections, are intended to address concerns about possible contaminated food imports. CBP is responsible for monitoring goods and materials in cargo shipments coming into the United States at all U.S. ports of entry, and is a regular part of inspection procedures carried out at every port of entry nationwide.[128] CBP's border inspections are intended to address concerns about possible contaminated food imports and are not specifically designed to address intentional contamination of food and food ingredients. CBP agriculture specialists prevent the entry of harmful plant and animal pests and diseases and confront emerging threats in agro- and bioterrorism, as well as ensure that the required permits, sanitary certificates (for animal products), and phytosanitary certificates (for plant products) accompany each product shipment.[129] As part of its role in enforcing plant and animal regulations, CBP has the authority to detain, where necessary, imported or exported products pending their clearance by agency inspectors. CBP was created in 2003 through a merger of the former U.S. Customs Service and the agricultural inspection portion of USDA's Animal and Plant Health Inspection Service (APHIS).[130]

Imported products must meet the same standards as domestic goods, and must contain informative and truthful labeling in English. Existing U.S. trade laws, such as general requirements under the Tariff Act of 1930 (19 U.S.C. § 1304), require all imported articles to be marked with the English name of the country of origin.[131] Other labeling requirements also apply under other laws that govern both FDA and USDA. For example, FDA requirements under FFDCA require that a food label must contain specified information. However, as noted by FDA, "The law does not specifically require that the country of origin statement be placed on the PDP [the principal display panel, PDP, or the label panel], but requires that it be conspicuous."[132] Certain labeling requirements for meat and poultry products are also required within laws administered by FSIS.[133] Only plants in countries certified by USDA to have inspection systems equivalent to those in this country are eligible to export products to the United States. Regulations require that country of origin appear in English on containers of all meat and poultry products entering the United States.[134] Other USDA-administered programs also provide for additional country-of-origin requirements for certain types of foods.[135]

Other Federal Food Quality or Food Safety Programs

A number of other federal agencies are involved in various food quality and food safety programs. Although these programs are not primarily focused on food fraud, product inspections under these programs might be leveraged to broadly address fraud concerns. In some cases concerns related to fraud are often a component of these programs. In other cases, such as in USDA's program certifying products that are organically produced, enforcement against fraudulent documents and certification is central to the program. Increased inspection and oversight as part of a program's product quality and marketing grades and standards may act as an added deterrent or improve detection if fraudulent marketing or related activities are happening.

USDA, Agricultural Marketing Service

Product Quality, Grading, and Standards

AMS is primarily responsible for product quality and marketing grades and standards for a range of foods, including dairy products, fruits and vegetables, livestock, meat, poultry, seafood, and shell eggs. However, AMS administers the egg surveillance program that inspects egg facilities quarterly to insure that egg handlers maintain required records and properly dispose of restricted eggs.[136] AMS also certifies quality programs and conducts quality grading services, generally user fee-funded. These programs are generally voluntary in nature, and in most cases do not directly address adulteration of food and food ingredients.

USDA programs establishing quality grade standards to encourage uniformity and consistency in commercial practices are provided for under the Agricultural Marketing Act of 1946 (7 U.S.C. §1621).[137] AMS develops quality grade standards for commodities as needed by the agriculture and food industry for a range of products, including cotton; dairy products; fresh and processed fruits and vegetables (and fruits and vegetables for processing); nuts and other specialty crops; livestock (including wool and mohair); poultry and eggs (including rabbits); and tobacco.[138] Under federal-state agreements, AMS-licensed state employees work where needed: in fields during harvest; at land, sea, and air ports of entry; and at packing houses, processing plants, warehouses, and federal and federal-state terminal markets. Grading is paid for by user fees and is voluntary unless the commodity is regulated for quality under a marketing order or agreement, subject to export requirements, or purchased by USDA or another federal agency for distribution (e.g., through the school lunch program or the military). Shipments of any imported commodity whose domestic production is under a marketing order or agreement must receive AMS grading to assure that the produce is comparable to the U.S. grade, size, quality, and maturity requirements.

Certified Organically Produced

AMS also oversees and enforces the USDA organic certification program, the National Organic Program (NOP). NOP is authorized under the Organic Foods Production Act of 1990, and the program's labeling and certification requirements are enforceable, and address concerns about fraud through product mislabeling.[139]

NOP regulations require that agricultural products labeled as "organic" originate from farms or handling operations certified by a state or private entity that has been accredited by USDA. Administered by AMS, NOP is a regulatory program that became operational in 2002, establishing a voluntary production and handling certification program. The program specifies the methods, practices, and materials that may be used and how certified organic production is to be grown, raised, and processed.[140] A central part of the NOP's stated mission is to "ensure the integrity" of USDA-certified organic products according to national standards for organically produced agricultural products that "assure consumers that products with the USDA organic seal meet consistent, uniform standards."[141] Since there are no tests that prove whether a food or ingredient is organically produced, certification relies on having the proper paperwork showing that processes were followed. Using fraudulent documents or certification to market, label, or sell non-organic (conventionally produced) food and food ingredients as USDA-certified "organic" is a violation of U.S. law and federal National Organic Program (NOP) regulations, punishable by fines of up to $11,000 for each violation.[142]

NOAA, National Marine Fisheries Service

NOAA's National Marine Fisheries Service (NMFS) administers a number of seafood and fisheries safety and sanitation programs. NMFS's voluntary seafood and fisheries safety inspection program focuses on marketing and product quality under the authority of the Agricultural Marketing Act of 1946 (7 U.S.C. §1621 *et seq.*).[143] The program offers additional levels and types of inspection that exceed FDA requirements, which program participants also must meet. Examples include onsite NOAA inspections during production hours, certification that plants or vessels meet specified sanitation requirements, quality inspections of individual product lots, and/or laboratory testing of products, among other services. NMFS works with FDA, which helps provide training and other technical assistance to NMFS. As part of its guidance, NOAA identifies common seafood fraud consisting of the addition of water or ice to add weight to the product; use of masking agents (such as carbon monoxide in tuna) that may give the fish added color or make it seem much fresher than it actually is; and also seafood substitution or labeling and selling a less expensive fish product as a more expensive product.[144]

Under the program, NMFS inspects a reported 20% of the seafood consumed in the United States.[145] Industry generally contracts with NMFS to provide the service, and NMFS personnel may inspect fishing vessels and processing plants to ensure that sanitary practices are in keeping with FDA standards. These services are provided on a fee-for-service basis and entitle participants to use various official grading and labeling marks, which are viewed as making their products more attractive to buyers. Exporters are often users of these services, in part because of foreign buyer requirements. NMFS may also periodically evaluate products at processing facilities for general condition, wholesomeness, and proper grading and labeling; and they may sample products for chemical and microbiological contamination, decomposition, and species identification.[146]

In addition, NOAA works with FDA and other federal agencies, as well as various state agencies, under the National Shellfish Sanitation Program (NSSP). NSSP is a federal/state cooperative program recognized by FDA and the Interstate Shellfish Sanitation Conference (ISSC) to promote and improve the sanitation of shellfish—oysters, clams, mussels, and scallops—moving in interstate commerce through federal/state cooperation and to promote uniformity of state shellfish programs.[147] Participants include agencies from states, FDA, EPA, NOAA, and the shellfish industry, and also foreign governments. Such cooperative efforts may act as a further deterrent to fraudulent activities, or improve detection if fraud is occurring.

Congressional Actions Involving Food Fraud

Congress has introduced a number of bills intended to address concerns about food fraud, mostly with respect to concerns about a particular food or food ingredient, but has not introduced legislation that would specifically address fraud in a comprehensive manner.

Most previous legislation in the past few years has tried to address fish and seafood fraud mostly through improved inter-agency cooperation and coordination. These include H.R. 1012/S. 520 (Markey/Begich) in the 113th Congress, S. 50 (Inouye) in the 112th Congress, and S. 3928 (Inouye) in the 111th Congress. Increased inspections of foreign seafood facilities were proposed in S. 2934 (Vitter) in the 111th Congress. Other previously introduced legislation would address fraudulent maple syrup (for example, H.R. 3363 in the 112th Congress).

Among other legislation introduced in the 113th Congress is H.R. 2400 (Capps), which seeks to improve recordkeeping and authorize investigations and enforcement actions for violations of the organic products standards. Previous congressional efforts have also highlighted concerns about fraudulently labeled "organic" products that might undermine the USDA-certified organic food industry.[148]

In addition, both the House-passed and Senate-passed versions of the 2013 farm bill (H.R. 2642/S. 954) include a provision to require USDA to submit a report to FDA that describes an appropriate federal standard for the identity of honey. This provision relates to previous congressional effort to push FDA to create "pure honey" standards that would allow federal border agents to better combat adulteration, misbranding, and fraudulent mislabeling of honey.[149]

A previous version of the House farm bill had also contained a provision that would establish tighter import controls on olive oil imports to enforce quality standards under the Agricultural Adjustment Act (7 U.S.C. §608e-1[a]).[150] This provision was removed by amendment during floor debate (H.Amdt. 213).

End Notes

[1] See, for example, B. Wilson, Swindled: From Poison Sweets to Counterfeit Coffee—The Dark History of Food Fraud, John Murray (Publishers), 2008; T. Mueller, Extra Virginity: The Sublime and Scandalous World of Olive Oil, W. W. Norton & Company, 2011; and S. Foster, "A Brief History of Adulteration of Herbs, Spices, and Botanical Drugs," HerbalGram, Issue 92 (pp. 42-57), 2011, American Botanical Council.

[2] CRS communication with researchers at the United States Pharmacopeial Convention (USP), December 23, 2013.

[3] For other background information, see CRS Report RL34080, Food and Agricultural Imports from China.

[4] H. Bottemiller, "Chinese Authorities Seize Melamine-tainted Dairy," Food Safety News, April 29, 2011.

[5] J. Moore, "The USP Food Fraud Database, and Beyond" Presentation at the USP Workshop of Economically Motivated Adulteration of Food Ingredients and Dietary Supplements, September 26–27, 2013.

[6] For other background information, CRS Report RL34124, Seafood Marketing: Combating Fraud and Deception.

[7] Department of Justice press release, "Former Officials and Broker of Peanut Corporation of America Indicted Related to Salmonella-Tainted Peanut Products," February 21, 2013.

[8] For other background information, see CRS Report R40450, Penalties Under the Federal Food, Drug, and Cosmetic Act (FFDCA) That May Pertain to Adulterated Peanut Products.

[9] See, for example, Import Alert # 36-03 (Detention Without Physical Examination of Honey Due to Chloramphenicol), June 20, 2013, http://www.accessdata.fda.gov /cms_ia /importalert_110.html.

[10] A. Schneider, "Tests Show most Store Honey Isn't Honey," Food Safety News, November 7, 2011.

[11] See, for example, FDA, "Patulin in Apple Juice, Apple Juice Concentrates and Apple Juice Products," September 2001; and "China to investigate rotten fruit juice," China.org,cn, September 24, 2013.

[12] USP press release, "Food Fraud Reports Up 60% Since 2010," January 23, 2013.

[13] European Commission, "Q&A on Horsemeat," http://ec.europa.eu/food and Ireland's Department of Agriculture, Food and the Marine, "Equine DNA and Mislabeling of Processed Beef Investigation," March 2013.

[14] Phenylbutazone is a non-steroidal anti-inflammatory medicine to treat musculoskeletal disorders, such as rheumatoid and arthritis, and is authorized for medicinal use in horses that are not intended for human consumption.

[15] GMA, Consumer Product Fraud, Deterrence and Detection, 2010, http://www.gmaonline.org /downloads/wygwam/ consumerproductfraud.pdf; and A. Kircher, NCFPD, "Tools for Protecting the Nation's Food Supply," June 5, 2012.

[16] Estimated by the Food Standards Agency of the U.K., as reported by K. Everstine and A. Kircher, "The Implications of Food Fraud," Food Quality & Safety magazine, June/July 2013.

[17] J. Spink and Z.L. Fejes, "A review of the economic impact of counterfeiting and piracy methodologies and assessment of currently utilized estimates," International Journal of Comparative and Applied Criminal Justice, September 2012.

[18] NCFPD, "EMA Frequently Asked Questions," http://www.ncfpd.umn.edu/.

[19] Comments attributed to John Spink, Michigan State University's Food Fraud Initiative, as reported by M. Oaklander, "11 Most Fraudulent Foods," Prevention.com.

[20] 74 Federal Register 64: 15497-15499, April 6, 2009.

[21] European Parliament, Committee on the Environment, Public Health and Food Safety draft report "on the food crisis, fraud in the food chain and the control thereof," (2013/2091(INI)).

[22] Regulation 178/2002 on general principles and requirements regarding food labeling, advertising, presentation, and packaging.

[23] FSA, "Food Fraud", http://www.food

[24] Cited examples include the recycling of animal by-products back into the food chain packing and selling of beef and poultry with an unknown origin knowingly selling goods which are past their "use by" date.

[25] Cited examples include farmed salmon sold as wild, and Basmati rice adulterated with cheaper varieties making false statements about the source of ingredients.

[26] J. Spink and in D.C. Moyer, Backgrounder: Defining the Public Health Threat of Food Fraud, National Center for Food Protection and Defense, April 2011.

[27] USP, Food Fraud Database, Glossary of Terms, http://www.foodfraud.org/glossary-terms.

[28] "Food Ingredients Most Prone to Fraudulent Economically Motivated Adulteration," Science Daily, April 5, 2012. Unconventional adulterants may trigger food safety concerns, as they might not be approved for use in food.

[29] NCFPD, "EMA Frequently Asked Questions," http://www.ncfpd.umn.edu.

[30] GMA, Consumer Product Fraud, Deterrence and Detection, 2010.

[31] J.C. Moore, J. Spink, and M. Lipp, "Development and Application of a Database of Food Ingredient Fraud and Economically Motivated Adulteration from 1980 to 2010," Journal of Food Science, Vol. 77, Nr. 4, 2012; and USP, Food Fraud Database, Glossary of Terms, http://www.foodfraud.org/glossary-terms.

[32] Based on work published in J. Spink and D.C. Moyer, "Defining the Public Health Threat of Food Fraud," Journal of Food Science, 2011, and other resources from MSU's Food Fraud Initiative (http://foodfraud.msu.edu/).

[33] CRS communication with Karen Everstine, NCFPD, December 27, 2013. For additional information on food defense, see section titled "Food Defense" in CRS Report R42985, Issues in Homeland Security Policy for the 113th Congress; and CRS Report RL34160, The National Bio- and Agro-Defense Facility: Issues for Congress. For more direct assistance, contact Sarah A. Lister, Specialist in Public Health and Epidemiology (slister@crs.loc.gov, 7-7320) or Jim Monke, Specialist in Agricultural Policy (jmonke@crs.loc.gov, 7-9664).

[34] NCFPD, "EMA Frequently Asked Questions," http://www.ncfpd.umn.edu.

[35] For example, presentations and comments at the USP Workshop of Economically Motivated Adulteration of Food Ingredients and Dietary Supplements, Rockville, MD, September 26–27, 2013.

[36] European Parliament, Committee on the Environment, Public Health and Food Safety draft report "on the food crisis, fraud in the food chain and the control thereof" (2013/2091[INI]).

[37] For general information on dietary supplements, see CRS Report R43062, Regulation of Dietary Supplements, or contact Amalia K. Corby-Edwards, Analyst in Public Health and Epidemiology (acorbyedwards@crs.loc.gov, 7-0423).

[38] V.M. Wheatley and J. Spink, "Defining the Public Health Threat of Dietary Supplement Fraud," Comprehensive Reviews in Food Science and Food Safety, Vol.12, 2013.

[39] D.S. Fabricant, "Dietary Supplement Adulteration". Presentation and comments at the USP Workshop of Economically Motivated Adulteration of Food Ingredients and Dietary Supplements, September 26–27, 2013.

[40] Comments by J. Spink, MSU, November 12, 2013, as part of the massive open online course (MOOC) on food fraud.

[41] USP Food Fraud Database, http://www.foodfraud.org/, and the USP, Food Chemicals Codex, 8th Edition. For information: http://www.foodfraud.org/frequently-asked-questions. USP is a "scientific nonprofit organization that sets standards for the identity, strength, quality, and purity of medicines, food ingredients, and dietary supplements manufactured, distributed and consumed worldwide" (http://www.usp.org/about-usp). Other information is at J. Moore, "The USP Food Fraud Database." Presentation at the USP Workshop of Economically Motivated Adulteration of Food Ingredients and Dietary Supplements, September 26–27, 2013.

[42] Detection technologies recorded in the database include high-performance liquid chromatography; standards for infrared spectroscopy; gas chromatography; isotope ratio mass spectrometry; hyphenated mass spectrometry methods; near infrared spectroscopy;

polymerase chain reaction; capillary electrophoresis; enzyme linked immunosorbent assay; thin-layer chromatography; site-specific natural isotope fractionation; mid-infrared spectroscopy; Raman spectroscopy; nuclear magnetic resonance spectroscopy; high-performance anion exchange chromatography; resonance spectroscopy; high-performance anion exchange chromatography; and differential scanning calorimetry.

[43] Freely available online at http://www.foodfraud.org/.

[44] J.C. Moore, J. Spink, and M. Lipp, "Development and Application of a Database of Food Ingredient Fraud and Economically Motivated Adulteration from 1980 to 2010," Journal of Food Science, Vol. 77, Nr. 4, 2012.

[45] Comments by J. Spink, MSU, November 12, 2013, as part of the massive open online course (MOOC) on food fraud. See also NCFPD, Final Report: 2007-2011, DHS Award #2007-ST-061-000003, p. 132.

[46] Information in the NCFPD EMA Incident Database is accessible via FoodSHIELD by request. For more information, see https://www.foodshield.org/index.cfm/join-registration /membership

[47] As a Homeland Security Center of Excellence, NCFPD is a multidisciplinary and action-oriented research consortium. For more information, see DHS, "Science & Technology Directorate Centers of Excellence," https://www.dhs.gov/st-centers-excellence; and NCFPD, "NCFPD's Mission: Defending the Safety of the Food System Through Research and Education," http://www.ncfpd.umn.edu/default/assets

[48] NCFPD, "NCFPD EMA Incidents Database," http://ema.foodshield.org/ (may require request for access). Also see NCFPD news release, "Preventing Economically Motivated Adulteration," February 4, 2013. Cited sources include LexisNexis, PubMed, Google, FDA consumer and recall records, state reports, and the E.U. Rapid Alert System for Food and Feed (RASFF) portal.

[49] A. Kircher, NCFPD, "Tools for Protecting the Nation's Food Supply," June 5, 2012; and information compiled by CRS from records in the NCFPD EMA Incident Database (database accessed November 14, 2013).

[50] "Port shopping" refers to when exporters and importers choose a particular port on the basis of their assessment of Customs' treatment, rather than on the quality of physical facilities and efficiency.

[51] K. Everstine, J. Spink, and S. Kennedy, "Economically Motivated Adulteration (EMA) of Food: Common Characteristics of EMA Incidents," Journal of Food Protection, Vol. 76, Nr. 4, April 2013. The analyses in this article covered 137 EMA incidents.

[52] Comments by Amy Kircher, NCFPD, USP Workshop of Economically Motivated Adulteration of Food Ingredients and Dietary Supplements, September 26, 2013; and NCFPD, "EMA Frequently Asked Questions."

[53] Additional information and the database is available at http://www.foodshield.org.

[54] DHS, "National Infrastructure Protection Plan: Food and Agriculture Sector," and DHS, Food and Agriculture Sector-Specific Plan: An Annex to the National Infrastructure Protection Plan, 2010.

[55] J.C. Moore, J. Spink, and M. Lipp, "Development and Application of a Database of Food Ingredient Fraud and Economically Motivated Adulteration from 1980 to 2010," Journal of Food Science, Vol. 77, Nr. 4, 2012.

[56] Natural flavoring complexes refer to essential oils, oleoresin, essence or extractive, which contains the flavoring constituents derived from a spice, fruit or fruit juice, vegetable or vegetable juice, edible yeast, herb, bark, bud, root, leaf or similar plant material, meat,

seafood, poultry, eggs, dairy products, or fermentation products used as a food ingredient for flavoring.

[57] FDA defines clouding agents as food processing aids "to enhance the appeal or utility of a food or food component, including clarifying agents ... , catalysts, flocculents, filter aids, and crystallization inhibitors" (21 CFR Part 1700).

[58] USP, "Food Fraud Reports Up 60% Since 2010," January 23, 2013.

[59] Oceana, "Oceana Study Reveals Seafood Fraud Nationwide," February 2013 (including previous reports in 2011); and Consumers Union, "Mystery Fish," Consumer Reports Magazine, December 2011. Similar investigations were also reported by the Boston Globe newspaper in 2011, http://www.boston.com/business/specials/fish

[60] FDA's Bad Bug Book: Foodborne Pathogenic Microorganisms and Natural Toxins Handbook, http://www.fda.gov/ Food/FoodborneIllnessContaminants/CausesOfIllnessBadBugBook/ucm071191.htm. See also K.A. Feldman et al., "A large outbreak of scombroid fish poisoning associated with eating escolar fish (Lepidocybium flavobrunneum)," Epidemiol Infect., 133(1):29-33, February 2005.

[61] See, for example, J. Field, "Puffer fish poisoning," J Accid Emerg Med. 15(5): 334–336, September 1998; and G.K. Isbister, M.C. Kiernan, "Neurotoxic marine poisoning," Lancet Neurol. 4(4):219-28, April 2005.

[62] See, for example, J.A. Tickner, et al., "Health risks posed by use of Di-2-ethylhexyl phthalate (DEHP) in PVC medical devices: a critical review," Am J Ind Med., 39(1):100-11, January 2001; and T.H. Yen et al., "Food safety involving ingestion of foods and beverages prepared with phthalate-plasticizer-containing clouding agents," J Formos Med Assoc., 110(11):671-84, November 2011.

[63] USP, "Food Fraud Reports Up 60% Since 2010," January 23, 2013.

[64] Information compiled by CRS from records in the NCFPD EMA Incident Database (database accessed November 14, 2013). Accessible to FoodSHIELD.org members.

[65] K. Everstine, J. Spink, and S. Kennedy, "Economically Motivated Adulteration (EMA) of Food: Common Characteristics of EMA Incidents," Journal of Food Protection, Vol. 76, Nr. 4, April 2013. The analyses in this article covered 137 EMA incidents.

[66] NCFPD news release, "Preventing Economically Motivated Adulteration," February 4, 2013.

[67] Communication between CRS personnel and researchers at University of Minnesota, November 27, 2013.

[68] European Parliament, Committee on the Environment, Public Health and Food Safety draft report "on the food crisis, fraud in the food chain and the control thereof," (2013/ 2091(INI)).

[69] Jurisdiction issues are addressed in a Memorandum of Understanding (MOU) between FSIS and FDA (MOU 225- 99-2001), http://www.fda.gov/AboutFDA/Partnerships Collaborations/MemorandaofUnderstandingMOUs/ DomesticMOUs/ucm117094.htm. Other information is in CRS Report RS22600, The Federal Food Safety System: A Primer.

[70] "A Guide to federal Food Labeling Requirements for meat, Poultry, and Egg Products," prepared for USDA Food Safety and Inspection Service by Hogan & Hartson, LLP, August, 2007, http://www.fsis.usda.gov/wps/wcm/connect/ f4af7c74-2b9f-4484-bb16-fd8f9820012d /Labeling_Requirements_Guide.pdf?MOD=AJPERES.

[71] See, for example, more recent cases involving olive oil and honey as described in A. Healy, "Lawsuit over olive oil labeling will go to trial," Food Chemical News, November 22, 2013; and DOJ press release, "Texas Honey Broker Sentenced to Three Years in Prison for Avoiding $37.9 Million in Tariffs on Chinese-Origin Honey," November 14, 2013.

[72] DOJ press release, "Former Officials and Broker of Peanut Corporation of America Indicted Related to Salmonella-Tainted Peanut Products," February 21, 2013.

[73] See, for example, National Oceanic and Atmospheric Administration (NOAA) press release, "NOAA Investigations Into Mislabeling Seafood Protects Consumers and Fishermen," February 4, 2011.

[74] Comments by J. Spink, MSU, November 12, 2013, as part of the massive open online course (MOOC) on food fraud.

[75] For a listing of programs, such as Food Defense Plan Builder and Food Related Emergency Exercise Bundle (FREE-B), see FDA, "Tools and Educational Materials," http://www.fda.gov/Food/FoodDefense/ToolsEducationalMaterials/.

[76] GAO, Fruit Juice Adulteration, Detection is Difficult and Enhanced Efforts Would be Costly, GAO/RCED-96-18, November 1995.

[77] GAO, Seafood Fraud, FDA Program Changes and Better Collaboration among Key Federal Agencies Could Improve Detection and Prevention, GAO-09-258, February 2009.

[78] GAO, Better Coordination Could Enhance Efforts to Address Economic Adulteration and Protect the Public Health, GAO-12-46, October 2011.

[79] USITC, Olive Oil: Conditions of Competition between U.S. and Major Foreign Supplier Industries," Inv. 332-537, August 2013.

[80] European Parliament, Committee on the Environment, Public Health and Food Safety draft report "on the food crisis, fraud in the food chain and the control thereof," (2013/2091(INI)).

[81] Estimated by backing out the reported 10%-20% of foods under USDA's jurisdiction. The 20% estimate is based on information reported by GAO in "Revamping Oversight of Food Safety," prepared for the 2009 Congressional and Presidential Transition, and appears to represent proportions of total spending for food consumed at home. The 10% estimate is based on data from USDA-ERS on U.S. per capita food consumption at http://www.ers.usda.gov/data/ foodconsumption/. See also DHS, "National Infrastructure Protection Plan: Agriculture and Food Sector Snapshot," http://www.dhs.gov/food-and-agriculture

[82] FSIS was authorized to inspect farmed catfish products under a 2008 farm bill provision (P.L. 110-246, §11016), but has not yet been implemented.

[83] MOU 225-99-2001, http://www.fda.gov/AboutFDA/PartnershipsCollaborations/MemorandaofUnderstandingMOUs/ DomesticMOUs/ucm117094.htm.

[84] See, for example, R. Carvajal, "USP, Food Fraud, and FSMA, FDA Law Blog, April 16, 2012, http://www.fdalawblog.net/fda_law_blog_hyman_phelps/2012/04/usp-food

[85] FFDCA §402(b) [21 U.S.C. §342(b)].

[86] P.L. 107-188.

[87] Owners, operators, or agents in charge of domestic or foreign facilities that manufacture/process, pack, or hold food for consumption in the U.S. are required to register the facility with FDA. See FDA, "Registration of Food Facilities," http://www.fda.gov/Food/GuidanceComplianceRegulatoryInformation/RegistrationofFoodFacilities/default.htm.

[88] FDA, "Registration Statistics," http://www.fda.gov/food/guidanceregulation/food facility registration/ucm236512.htm.

[89] Regulations are 21 CFR Parts 130-169. 21 CFR 130 covers general requirements.

[90] For an informative overview of how EMA activities have evolved FDA, see J. Spink, "Economically Motivated Adulteration: Another Dimension of the 'Expanding Umbrella of Food Defense,'" Food Safety Magazine, October/November 2013.

[91] ORA is the lead office for all agency field activities, and inspects regulated products and manufacturers, conducts sample analyses of regulated products and reviews imported products offered for entry into the United States.

[92] J. Thomas, "FDA Workgroup on Economically Motivated Adulteration," Presentation at the USP Workshop of Economically Motivated Adulteration of Food Ingredients and Dietary Supplements, September 26–27, 2013.

[93] GAO, Better Coordination Could Enhance Efforts to Address Economic Adulteration and Protect the Public Health, October 2011, GAO-12-46.

[94] J. Murphy, "FDA Struggles with FSMA-Required Role in Combating Food Fraud," Food Chemical News, May 10, 2013.

[95] P.L. 111-353. For information, see CRS Report R40443, The FDA Food Safety Modernization Act (P.L. 111-353).

[96] FDA's rulemaking documents are available at http://www.fda.gov/Food/Guidance Regulation/FSMA/ ucm253380.htm.

[97] See, for example, R. Carvajal, "USP, Food Fraud, and FSMA, FDA Law Blog, April 16, 2012, http://www.fdalawblog.net/fda_law_blog_hyman_phelps/2012/04/usp-food

[98] For more information, see FDA, "FSMA Proposed Rule for Preventive Controls for Human Food Current Good Manufacturing Practice and Hazard Analysis and Risk-Based Preventive Controls for Human Food," http://www.fda.gov/Food/GuidanceRegulation /FSMA/ucm334115.htm (docket#: FDA-2011-N-0920).

[99] 78 Federal Register 11: 3695. January 16, 2013.

[100] HACCP refers to "Hazard Analysis & Critical Control Points" and is defined by FDA as a management system to address food safety through the analysis and control of biological, chemical, and physical hazards from raw material production, procurement and handling, to manufacturing, distribution and consumption of the finished product.

[101] FDA, "Frequently Asked Questions," http://www.fda.gov/Food/GuidanceRegulation /FSMA/ucm247559.htm.

[102] For more information, see FDA, "Focused Mitigation Strategies to Protect Food Against Intentional Adulteration," December 2013, http://www.fda.gov/Food/Guidance Regulation/ FSMA/ucm378628.htm.

[103] For more information, see FDA, FSMA Reports and Studies, http://www.fda.gov/Food /GuidanceRegulation/FSMA/ ucm271961.htm.

[104] For more information, see J. Spink, C.T. Elliott, and K.P. Swoffer, "Defining Food Fraud Prevention to Align Food Science and Technology Resources," Food Science & Technology Journal, Vol. 27, Nr. 4, December 2013.

[105] For more information, see FDA, "Food Supplier Verification Programs (FSVP) for Importers of Food for Humans and Animals," July 2013 (docket# FDA-2011-N-01436); and "Accreditation of Third-Party Auditors/Certification Bodies to Conduct Food Safety Audits and to Issue Certifications," July 2013 (docket# FDA-2011-N-01468); and "Information Required in Prior Notice of Imported Food," May 2013 (docket# FDA-2011-N-017911), available at http://www.fda.gov/Food/GuidanceRegulation/FSMA/ucm253380.htm.

[106] Reports required under FSMA are available at http://www.fda.gov/food

[107] FDA, "2012 Annual Report on Food Facilities, Food Imports, and FDA Foreign Offices," August 2012, http://www.fda.gov/Food/GuidanceRegulation/FSMA/ucm315486.htm

[108] Comments attributed to FDA Deputy Commissioner for Foods and Veterinary Medicine, Mike Taylor, reported by J. Murphy, "FDA Struggles with FSMA-Required Role in Combating Food Fraud," Food Chemical News, May 10, 2013.

[109] FDA commissions CBP to assist FDA with examinations and investigations related to prior notice requirements for imported goods (21 U.S.C. §381(m)) at ports and other facilities and locations subject to CBP jurisdiction, under MOU 225-04-4001 (http://www.fda.gov /AboutFDA/PartnershipsCollaborations/MemorandaofUnderstandingMOUs/ DomesticMOUs/ucm115145.htm).

[110] If a shipment is refused admission, the importer may introduce evidence within 10 days to avoid the appearance of a violation. During that time, the product is held at a warehouse or with the importer and cannot be distributed. If the shipment is not proven to be safe, it must be destroyed or exported within 90 days.

[111] FDA, "Import Alerts Guard Against Unsafe Products," August 25, 2011. The database is at http://www.fda.gov/ Forindustry/ImportProgram/ImportAlerts/default.htm.

[112] FDA, "Detention without Physical Examination," http://www.fda.gov/ICECI/Compliance Manuals/ RegulatoryProceduresManual/ucm179271.htm; and Linda Horton, "U.S. FDA Authority over Imports," http://www.hoganlovells.com/files/Publication/a279dfd1-c81e-464e-a4f0-b26b6da16f29/Presentation/ PublicationAttachment/4511545b-3ab8-4982-a2f6-ccfa2990c0a8/Horton.pdf.

[113] An entry line refers to a unique shipment.

[114] FDA, "2012 Annual Report on Food Facilities, Food Imports, and FDA Foreign Offices," August 2012, http://www.fda.gov/Food/GuidanceRegulation/FSMA/ucm315486.htm

[115] GAO, Fundamental Changes Needed to Ensure Safe Food (GAO-02-47T), October 10, 2001, http://www.gao.gov/ new.items/d0247t.pdf.

[116] For more information, see CRS Report RL32922, Meat and Poultry Inspection: Background and Selected Issues. For more direct assistance, contact Joel L. Greene, Analyst in Agricultural Policy (jgreene@crs.loc.gov, 7-9877).

[117] MOU 225-99-2001, http://www.fda.gov/AboutFDA/PartnershipsCollaborations/ Memoranda of UnderstandingMOUs/DomesticMOUs/ucm117094.htm.

[118] "A Guide to federal Food Labeling Requirements for meat, Poultry, and Egg Products," prepared for USDA Food Safety and Inspection Service by Hogan & Hartson, LLP, August, 2007, http://www.fsis.usda.gov/wps/wcm/connect/ f4af7c74-2b9f-4484-bb16-fd8f9820012d /Labeling_Requirements_Guide.pdf?MOD=AJPERES.

[119] P.L. 110-246, §11016 (Inspection and Grading). USDA has not yet finished its catfish inspection rule (76 Federal Register 10434, February 24, 2011), and will inspect catfish facilities when the rule is finalized.

[120] In a Hazard Analysis and Critical Control Point (HACCP) plan a facility must identify each point in its processes where contamination could occur ("critical control point") and have a plan to control it, as well as document and maintain records.

[121] For more information, see CRS Report RL32922, Meat and Poultry Inspection: Background and Selected Issues.

[122] For information, see FSIS, "FSIS Import Procedures for Meat, Poultry & Egg Products," http://www.fsis.usda.gov/.

[123] FSIS' inspection directory, http://www.fsis.usda.gov/wps/portal/fsis/topics/inspection/mpi-directory.

[124] FSIS regulations are at 9 CFR 300. For meat products, labeling regulations are at 9 CFR 317; for poultry, regulations are at 9 CFR 381. Regulations regarding weights are at 9 CFR 442. Other information on FSIS labeling requirements is at http://www.aamp.com /regulations/fsis-labeling/.

[125] FMIA (21 U.S.C. §601) and PPIA (21 U.S.C. §453).

[126] CRS inquiry (via FSIS at http://www.fsis.usda.gov/wps/portal/informational/askkaren) received December 17, 2013. Horses are not allowed to be slaughtered and horse meat is not allowed to be processed in the same facility as other species in the United States. FSIS notice on increased species sampling in response to horsemeat is at http://www. fsis.usda.gov/OPPDE/rdad/FSISNotices/25-13.pdf.

[127] Regulations are 9 CFR Part 319 for meat products and 9 CFR §§381.156 to 381.174 for poultry products. For corresponding labeling information, see FSIS' "Food Standards and Labeling Policy Book," August 2005, http://www.fsis.usda.gov/OPPDE/larc/Policies /Labeling_Policy_Book_082005.pdf.

[128] The authority to search, inspect, and examine merchandise entering the United States is in 19 U.S.C. §1467; regulations are at 19 C.F.R. 162.6. Other search authorities include 19 U.S.C. §§482, 1496, 1581 and 1582. These authorities rest with the U.S. Treasury that typically delegates its authority via DHS to CBP.

[129] USDA, "Importing Food and Agricultural Products into the United States," August 2012.

[130] Memorandum of Agreement (MOA), DHS Agreement# BTS-03-0001 and USDA-APHIS Agreement# 03-1001- 0382-MU, March 2003, http://www.aphis.usda.gov/plant_health

[131] Customs and Border Protection's Guidance is at http://www.fda.gov/ICECI/Compliance Manuals/ CompliancePolicyGuidanceManual/ucm074567.htm. See regulations at 19 CFR 134.

[132] 21 CFR 101. See FDA, "Guidance for Industry: A Food Labeling Guide," October 2009, http://www.fda.gov/Food/GuidanceComplianceRegulatoryInformation/GuidanceDocuments /FoodLabelingNutrition/FoodLabelingGuide/ ucm064872.htm.

[133] FMIA (21 U.S.C. §601 et seq.) and PPIA (21 U.S.C. §451 et seq.).

[134] Regulations are at 9 C.F.R. 327.14 and 9 C.F.R. 381.205.

[135] Includes Country of Origin Labeling (COOL); see CRS Report RS22955, Country-of-Origin Labeling for Foods and the WTO Trade Dispute on Meat Labeling; and Perishable Agricultural Commodities Act (PACA) of 1930 and the Produce Agency Act of 1937 (7 U.S.C. §499a et seq., and §1622, respectively); see CRS Report R42771, Fruits, Vegetables, and Other Specialty Crops: Selected Federal Programs.

[136] Egg Products Inspection Act, as amended (21 U.S.C. §§1031-1056), 7 CFR Part 57.

[137] 7 U.S.C. §§1621 et seq., 7 CFR Part 75.

[138] USDA, "USDA Quality Standards," http://www.ams.usda.gov/standards/.

[139] 7 U.S.C. §§6501-6522, 7 CFR Part 205. OFPA was enacted as part of the 1990 farm bill (P.L. 101-624).

[140] NOP regulations prohibit the use of genetic engineering, irradiation, and sewage sludge in certified organic production and handling.

[141] For information, see USDA's website (http://www.ams.usda.gov/AMSv1.0/nop). See also NOP's "Fact Sheet," NOP's "2010-2012 Strategic Plan," and "National Organic Program," http://www.ams.usda.gov/AMSv1.0/organicinfo.

[142] NOP's website lists prior and ongoing cases of fraudulent organic certificates at http://www.ams.usda.gov/nop.

[143] NOAA Seafood Inspection Program, http://www.seafood. See also CRS Report RS22797, Seafood Safety: Background and Issues. For more direct assistance, contact Harold F. Upton, Analyst in Natural Resources Policy (hupton@crs.loc.gov, 7-2264).

[144] See, for example, NOAA, "Identifying Seafood Fraud: A Common Practice with Serious Consequences," http://www.fishwatch.gov/buying_seafood

[145] NOAA, "Inspecting Seafood—A Highly Trained Nose Knows," Fishwatch.gov, http://www.fishwatch.gov/features/ steve_wilson_10_15_12.htm.

[146] FDA, "2012 Annual Report on Food Facilities, Food Imports, and FDA Foreign Offices," August 2012, http://www.fda.gov/Food/GuidanceRegulation/FSMA/ucm315486.htm.

[147] Information is at FDA, http://www.fda.gov/food. As part of a 1984 Memorandum of Understanding, ISSC was recognized as the primary voluntary national organization of state shellfish regulatory officials, providing guidance and counsel on matters for the sanitary control of shellfish.

[148] See, for example, letter from Senator Charles E. Schumer to USDA Secretary Tom Vilsack, September 20, 2010, available at http://www.schumer.senate.gov/Newsroom/record_print.cfm?id=327768.

[149] See, for example, Senator Schumer's press release on June 11, 2010, regarding a letter sent to FDA Commissioner Margaret Hamburg. See also J. Pecquet, "Schumer: 'Honey laundering' a sticky problem that needs FDA intervention," The Hill, June 12, 2010.

[150] H.R. 1947, §10010. See, for example, N. Flagg, "New Farm Bill Includes Olive Oil Controls," May 20, 2013.

In: Food Fraud and Adulterated Ingredients ISBN: 978-1-63117-730-9
Editor: Darrell T. Braden

Chapter 2

SEAFOOD MARKETING: COMBATING FRAUD AND DECEPTION*

Eugene H. Buck

SUMMARY

Congress is facing questions of whether the law applicable to fraudulent seafood sales and marketing is clear and enforceable, whether agency enforcement efforts targeting seafood fraud are adequate, and whether the penalties for seafood fraud are a deterrent. Congress may become involved in oversight of how federal agencies are addressing these issues, and legislation related to these concerns may be considered.

With increased seafood imports and decreased monitoring, fraud and deception in seafood marketing appears to be growing more widespread. The flesh of many fish species is similar in taste and texture and, therefore, it is difficult to identify species in fillet form, especially after preparation for consumption. Thus, it can be relatively easy to substitute an inexpensive species for one of higher value. Inaccurate (low) counts or net weights ("short weighting") result in consumers receiving less for their money than advertised and anticipated. Overbreading may cause consumers to pay shrimp prices for excess bread crumbs. Excessive amounts of ice glaze (overglazing) can deliberately be used to increase the apparent weight, and therefore the apparent value, of the delivered

* This is an edited, reformatted and augmented version of a Congressional Research Service publication, CRS Report for Congress RL34124, prepared for Members and Committees of Congress, from www.crs.gov, dated January 2, 2013.

product. In addition, some new treatment procedures by the seafood industry, such as carbon monoxide/tasteless smoke, are being questioned for their potential to deceive consumers. Since food safety and bioterrorism concerns have taken precedence, regulatory agencies have been less able to maintain control of economic fraud.

The extent of this fraud is not well documented. The National Fisheries Institute (NFI) has undertaken an initiative to promote economic integrity within the seafood industry, concentrating on three primary areas:

- transshipment of products to avoid antidumping and countervailing duties;
- mislabeling of products or species substitution; and
- mislabeling of weights or counts of products.

The Food and Drug Administration (FDA) is the primary agency responsible for ensuring that food sold in interstate commerce is properly labeled. FDA's jurisdiction covers seafood and the agency operates an oversight compliance program, the Seafood Regulatory Program, for fishery products. Responsibility for a food product's safety, wholesomeness, identity, and economic integrity rests with the processor or importer, who must comply with regulations promulgated under the Federal Food, Drug and Cosmetic Act (FFDCA) and the Fair Packaging and Labeling Act (FPLA).

In the 112th Congress, S. 50 proposed directing the Departments of Commerce and of Health and Human Services, the Federal Trade Commission, and other federal agencies to combat seafood fraud; on January 26, 2012, the Senate Committee on Commerce, Science, and Transportation reported this bill. H.R. 6200 sought to address seafood fraud by requiring labels to identify species and origin for both domestic and imported fish and by requiring a plan to coordinate FDA and National Marine Fisheries Service seafood inspection.

This report reviews recent incidents of fraud and deception and examines related policy issues.

INTRODUCTION

Media attention has focused on recent incidents of fraud relating to seafood—restaurants knowingly serving lower-priced fish than identified on menus, lower-priced species marketed commercially as higher-priced species, packaged weights of seafood being less than labeled weights, and extra water being added to seafood to increase total product weight—raising public concern.[1] Most seafood fraud is based on supplying the consumer with something different from and inferior to the product expected.[2] In some instances, such practices may not be intentional, and the extent of intentional fraud is not well documented. These occurrences have the potential to erode consumer confidence in seafood generally, potentially lowering Americans' consumption of seafood. On a broader scale, seafood fraud is recognized as one component of international concerns related to illegal, unreported, and unregulated (IUU) fish harvests.[3] Congress is facing questions of whether U.S. law applicable to fraudulent seafood sales and marketing is clear and enforceable, whether agency enforcement efforts targeting seafood fraud are adequate, and whether the penalties for seafood fraud are a deterrent.

The primary federal law that addresses mislabeling is the Federal Food, Drug, and Cosmetic Act of 1938 (FFDCA; 21 U.S.C. §§ 301 et seq.), which is administered by the Food and Drug Administration (FDA). The FFDCA, as amended, gives FDA authority over most food regulation and includes (1) definitions elaborating on the concepts of adulteration and misbranding; (2) provisions for control over all labeling of foods in interstate commerce; (3) detailed regulation of safety and wholesomeness of foods; and (4) enforcement remedies available to the agency. In addition, the Fair Packaging and Labeling Act (FPLA; 15 U.S.C. §§ 1451 et seq.) requires that consumers of packaged commodities be provided with accurate information as to its contents. Food products falling within the scope of the FFDCA that are introduced into interstate commerce in violation of the FPLA and its regulations are deemed to be misbranded within the meaning of the FFDCA. The Chemical Services Group of the National Marine Fisheries Service (National Oceanic and Atmospheric Administration, Department of Commerce) at the National Seafood Inspection Laboratory in Pascagoula, MS, provides analytical testing to identify fish species. Applicable law is discussed in detail below.

FRAUDULENT OR DECEPTIVE PRACTICES

Mislabeling or Substituting Species

The differences in the taste and texture of different fish species' flesh may be subtle, and therefore it is frequently difficult to identify a species in fillet form, especially after it is prepared for consumption. Thus, it is relatively easy to substitute an inexpensive species for one of higher value. Over the nine-year period of FY1988-FY1997,4 routine examinations of seafood products by the

National Marine Fisheries Service's National Seafood Inspection Laboratory (NSIL) found that 37% of fish and 13% of other seafood (e.g., shellfish, edible seaweed) from randomly selected vendors were mislabeled.[5]

Identification of fish flesh usually requires DNA testing. There is debate within the seafood laboratory testing community about the accuracy of public DNA sequence databases for fish. Public databases seldom use validated fish standards and thus are likely to contain incomplete and/or inaccurate DNA sequences. Although most regulatory agencies are developing DNA methods for species identification,[6] some of these agencies continue to depend on a method— isoelectric focusing electrophoresis of soluble muscle proteins—that is unreliable for species identification of heat-processed or dried fish products.[7] Some U.S. seafood importers and distributors employ private companies to conduct random testing to eliminate blatant species substitution and minimize unintentional species substitution. DNA methods for fish species identification are not yet officially recognized by FDA, but the agency is working with an international group, FISH-BOL,[8] to develop a qualified database of DNA sequences from taxonomically verified fish specimens.[9] This database is still being developed and is not yet available for use. However, FDA has purchased gene sequencing instrumentation for field laboratories and has conducted pilot testing of samples from various sources so that this new molecular technology can be used on a more routine basis.[10]

Fraud can be perpetuated in several ways. Unfair and deceptive trade practices occur when restaurants misrepresent menu items to their patrons by substituting other (often less desirable and less expensive) fish for an item described as a higher-valued species. Fraud also occurs at the manufacturing level, as in American Samoa, where six tuna cannery workers were convicted for their involvement in falsely labeling hundreds of cans as albacore tuna when the cans were actually filled with less expensive wahoo, and selling them to local stores.[11] Some distributors have knowingly sold restaurants and retailers lower-valued species, claiming that they are different species of higher value.

While species substitution may be intentional for certain species and products because of their differing values, it also occurs unintentionally where species identities are easily mistaken. *Table 1* provides a list of some common illegally substituted species. Additional instances of mislabeling include imported farmed salmon being falsely identified as wild Alaska salmon[12] and frozen seafood being marketed as fresh product, as well as confusion over seafood being marketed as "organic." In 2009, a Washington state man was fined $160,000 and sentenced to 30 days in jail for intentionally mislabeling

136,000 pounds of turbot from China as much higher priced U.S. halibut.[13] Since large quantities of seafood are imported for U.S. consumption, some portion of the mislabeling problem undoubtedly originates with foreign suppliers.[14] To address this, FDA has issued an import alert on species substitution, providing guidance to agency field personnel about the manufacturers and/or products at issue.[15] Asian catfish (basa and tra) may illustrate the magnitude of the problem. Although import volumes exceed 50 million pounds annually for these species combined, they are difficult to find correctly identified in the domestic marketplace.[16]

Table 1. Examples of Common Illegally Substituted Seafood (less expensive products are in column B)

A	B
Red snapper	Rockfish
Mahi mahi	Yellowtail (Seriola lalandi)
Swordfish	Mako shark
Orange roughy	Oreo dory or John dory
Cod	Alaska pollock
Halibut	Sea bass
Dover sole	Arrowtooth flounder
Red drum	Black drum
Snapper (Lutjanus sp.)	Tilapia
Grouper	Basa or tra
Lake or yellow perch	White perch or zander
Caviar (sturgeon species)	Paddlefish or other fish roe
Walleye	Sauger or Alaska pollock
Chum salmon	Pink salmon
Salmon	Steelhead trout
Pacific salmon	Atlantic salmon
Blue crab meat	Imported crab meat
Wild-caught salmon	Farm-raised salmon

Source: Table contents modified from http://www.cfsan.fda.gov/~frf/econ.html.

Problems can arise from substitution at the restaurant level, misrepresentation by the restaurant supplier, or product misidentification anywhere in the harvesting and processing system. It is often difficult to determine who may have been at fault, especially if there is collusion. Intentional fraud, rather than human error, is more likely to have occurred where "cheaper" species are consistently mislabeled as more "valuable" species, rather than the other way around. One recent study, based on an analysis of mitochondrial DNA sequences, revealed that 77% of the fish sold in the U.S. marketplace as red snapper, *Lutjanus campechanus*, belonged to other species of the Lutjanidae family, and could not legally be labeled as "red snapper."[17] In early 2007, Alabama agriculture inspectors targeted 35 Gulf of Mexico restaurants and seafood markets for sampling after a Mobile, AL, television station used DNA testing and found that only one in ten samples from restaurants advertised as grouper actually were this fish.[18] Alabama state law makes a restaurant or distributor subject to a fine of $5,000 per offense for selling falsely labeled seafood products. On September 3, 2008, the Florida Attorney General's Office announced a settlement with Sysco Food Services—West Coast Florida, Inc., requiring the distributer to refrain from marketing to its restaurant clients any items described as grouper without taking commercially reasonable steps to determine that the item is genuine.[19] This legal action was initiated after a similar media investigative report in Tampa, FL, discovered that 17 of 24 area restaurants appeared to be selling less expensive species as grouper.[20] Media investigations have not been limited to the Gulf area.[21] Because the laboratories that tested fish samples for media inquiries all used DNA sequencing methods and made fish species identifications based on DNA sequences in public databases, there is debate within the seafood scientific laboratory testing community about the accuracy of these media reports. A more recent study of New York City restaurants and markets used DNA barcode technology[22] to determine that 14 of 56 samples collected were incorrectly labeled as higher-priced fish.[23]

Common Names of Fish

The correct use of names is crucial for properly identifying seafood. Because one species may have different vernacular names in different regions and because different species may be called by the same vernacular name, standard market names for seafood products are needed to avoid confusion. Common names of fish species have been standardized, often by professional societies.[24] FDA and the National Marine Fisheries Service (NMFS) have cooperated to develop "The Seafood List," compiling existing acceptable market names for imported and domestically available seafood.[25] In 1988, FDA published the FDA Guide to Acceptable Market Names for Food Fish Sold in Interstate Commerce ("The Fish List") to provide an authoritative source of common names to establish order in the marketplace and reduce confusion among consumers. In 1993, FDA published an updated, expanded "Seafood List," which includes invertebrate species (mollusks and crustaceans) as well as finfish. The frequently updated Seafood List reflects what FDA considers the most appropriate market names for the identification and labeling of seafood and is the agency's primary guidance for naming seafood sold in interstate commerce.

Two main provisions cover the naming of fish—Section 403 of the Federal Food, Drug, and Cosmetic Act of 1938 (FFDCA), which addresses misbranded food, and 21 C.F.R. § 102.5, which sets forth general principles for common or usual names for non-standardized food. A common or usual name is the "prevalent and meaningful name by which consumers ordinarily identify the food."[26] Use of either the acceptable market name or the common name in labeling seafood products assures that identity labeling of the seafood will comply with FDA and National Marine Fisheries Service regulations. FDA discourages the use of vernacular names as this practice may cause seafood to be misbranded.[27]

An example of this concern arose several years ago when increasing imports of basa (*Pangasius bocourti*) from Vietnam were marketed in the United States as "catfish," causing confusion with domestically produced *Ictalurid* catfish. FDA guidance formerly listed a number of fish other than those from the family *Ictaluridae* with the term "catfish" in their names. This prior guidance reflected what FDA believed were names for seafood that could be used by importers, domestic distributors, and sellers consistent with the food naming provisions of the FFDCA. To address this confusion, § 10806 of P.L. 107-171 (Farm Security and Rural Investment Act of 2002) amended § 403 (the food misbranding provision) of the FFDCA (21 U.S.C. § 343) to provide that a food shall be deemed to be misbranded "[i]f it purports to be or is represented as catfish, unless it is fish classified within the family *Ictaluridae*." Although additional concern relates to the use of "lobster" in describing items generally identified as "langostinos,"[28] the extent of this problem is unknown.

Country-of-Origin Labeling

Mislabeling of the country of origin is another concern. Without labeling, consumers would rarely be able to distinguish a product's country of origin; labeling fish falsely as to country of origin removes the cachet from more desirable products, driving down the more desirable products' market price. Importers may falsely claim their seafood product is from a country where seafood products may be recognized as of higher quality than the actual source country.

Particularly for seafood, some difficult questions relate to the meaning of the term "country of origin." For example, is product origin best represented by where the seafood was caught, the flag of the harvesting vessel, where the harvest was first landed, where the product was first processed, or where it was last or secondarily processed? Following the terrorist attacks on the World Trade Center on September 11, 2001, interest in traceability in the food supply chain increased substantially, including recognition of the advantages of a complete traceability program for seafood from producer to customer so that the source of and liability for any mislabeling could be more easily identified.[29]

In October 2004, the Agricultural Marketing Service (AMS), U.S. Department of Agriculture (USDA), promulgated an interim final rule requiring certain retailers and their suppliers to notify customers of the country of origin of wild and farm-raised fish and shellfish.[30] Final rules to implement fully the country-of-origin requirements were published by AMS on January 15, 2009.[31] USDA has entered into agreements with states having existing enforcement infrastructure to assist in compliance reviews for fish and shellfish. Only USDA is able to initiate enforcement actions against a person found to be in violation of the law, as this statute does not provide for a private right of action. NOAA was reported to be conducting a criminal investigation of alleged repacking and relabeling of imported farmed and wild shrimp as domestic U.S. product.[32]

In April 2009, a Washington State man was fined $160,000 and sentenced to 30 days in jail for mislabeling 136,000 pounds of turbot from China as U.S. halibut.[33]

Low Weights or Undercounting

Inaccurate (low) counts or net weights ("short weighting") result in consumers receiving less for their money than advertised and anticipated.[34] These instances, although commonly reported, also constitute mislabeling offenses under the FFDCA. The seafood community recognizes this problem to be of greater concern than species substitution, costing legitimate businesses sales and reduced confidence in their true-packaged products. As part of its economic integrity initiative (see "Industry Initiatives," below), the National Fisheries Institute reported that, in the year preceding August 2008, it sent more than 80 letters to FDA documenting specific examples of fraudulent weights.[35]

Over-Treating or Added Water Weight

The FDA standard for breaded shrimp requires that such a product contain at least 50% shrimp. Overbreading may cause consumers to pay shrimp prices for excess bread crumbs. Frozen fillets, shrimp, crab legs, and other products are normally protected from dehydration (freezer burn) while frozen by the application of a light glaze of ice, and the water weight of this glaze is not included in the stated product weight. Excessive amounts of glaze (overglazing), not compensated for in this manner, can deliberately be used to increase the apparent weight, and therefore the apparent value, of the delivered product. On February 20, 2009, FDA reissued its 1991 guidance, warning the seafood industry that short weight on a product obtained by including glaze in stated weights is a felony.[36] Despite this warning, a multistate investigation in early 2010 found this problem to be widespread.[37] In response to this problem, the American Scallop Association, representing the majority of U.S. scallop production, announced a plan to enforce labeling and packing standards for its 18 members to combat short-weight scallop packs.[38]

Sodium tripolyphosphate (STPP) is used in the seafood industry to retain moisture in the product so as to prevent freezer burn and, when properly used, can be a legitimate means for aiding processing. However, STPP can be misused to retain excess moisture in seafood products. Prolonged soaking of seafood in an STPP-water solution can result, for example, in Atlantic sea scallops or shrimp with excessive water. Such excess water adds to the product's total weight, resulting in misrepresentation when seafood prices are charged for the extra water and, in the case of shrimp, the product is bumped into a larger weight class where a higher price per pound can be charged. Seafood treated with STPP or other water-retaining chemicals must be accurately labeled to identify this treatment. However, FDA has found it difficult to set percent-moisture guidelines and labeling requirements for treated scallops,[39] and no moisture guidelines or standards exist for shrimp. Some critics suggest that the use of all water retention chemicals (primarily phosphates) should be prohibited in fish and shellfish.

Altered Color

Fish fillets can be treated with carbon monoxide (CO) to give fish flesh a fresher-appearing reddish tint.[40] The growing use of CO (also referred to as "tasteless smoke" or TS) as a "pigment fixative" has alarmed some consumer advocates who say it deceives shoppers who depend on color to help them avoid spoiled fish. Seafood industry representatives assert that TS/CO is a legitimate preservative, and the FDA has said it has no questions regarding an industry determination that TS/CO is a substance generally recognized as safe.[41] An additional consumer safety issue can occur when the flesh of certain species such as tuna develops toxic levels of histamine through time and/or temperature abuse—with CO treatment there are no visual cues to indicate when such flesh may be decomposed and toxic. Consumer advocates have urged FDA to conduct a formal evaluation of this treatment's impact on consumer safety. FDA considers tuna to be misbranded if it is treated with TS/CO but not labeled to indicate that it contains a preservative and thus purports to be unprocessed, fresh, or fresh-frozen tuna.[42] Thus, all processed seafood items involving TS/CO require label declarations under 21 C.F.R. Part 101.22(j).[43] In the 111th Congress, language was introduced in H.R. 759 and H.R. 2749 that would have amended the FFDCA to restrict the use of CO in meat, poultry, and seafood. However, when H.R. 2749 was reported, this language was deleted.

Some aquaculture operations use the color additives canthaxanthin and/or astaxanthin in feed to impart a more orange color to fish flesh of salmon and/or trout.[44] The flesh of the farmed varieties of these fish would, if not for these color additives, be a possibly less appealing, paler color. Under specified conditions, use of these additives in feed for salmonid fish is legal as long as fish are properly labeled to identify that this treatment has been used.[45] Farmed salmon and trout, where additives have been used to enhance color, are susceptible to being improperly labeled and this product deceptively and fraudulently marketed as "wild" fish.[46]

Transshipment and Mislabeling to Avoid Customs Duties

Transshipment occurs when foreign producers ship goods through a second country en route to the United States. Although transshipment is generally legal and commonly used in the ordinary course of business, it is illegal if it is done for the purpose of circumventing duties and other applicable trade restrictions. It has been reported that shrimp from China have been shipped to the United States by way of Indonesia to avoid paying antidumping duties of 112% levied by the United States on shrimp imported from China, but not on shrimp imported from Indonesia.[47] In other cases, seafood such as Asian catfish has been mislabeled as sole specifically to avoid paying antidumping duties.[48]

In addition, it is a violation of the Lacey Act (16 U.S.C. 3372(d)) to falsely identify any fish, wildlife, or plant transported in interstate or foreign commerce.

INDUSTRY INITIATIVES

In 1986, the Southeastern Fisheries Association, funded under the Saltonstall-Kennedy Act grant program,[49] introduced a Seafood Product Quality Code program in one of the first efforts by a fisheries trade association to better inform wholesale and retail seafood buyers.[50] The code, an educational and promotional concept, cataloged the industry's recommended measures for product quality for individual species to better inform buyers and to encourage more industry quality compliance by facilitating communications between buyers and sellers. Compliance with the code was a voluntary means whereby seafood sellers could indicate that they provide consumers a very safe, high-quality product.

In October 2006, concerns that seafood fraud had begun to and could increasingly erode consumer confidence in seafood led the National Fisheries Institute (NFI) to announce an initiative to promote economic integrity within the seafood industry. Implementation of this initiative commenced in summer 2007, concentrating on three primary areas:

- transshipment of products to avoid antidumping and countervailing duties;
- mislabeling of products or species substitution; and
- mislabeling of weights or counts of products.

NFI officially launched its Better Seafood Bureau on July 5, 2007.[51] NFI is pursuing their initiative by (1) obtaining commitments from the CEOs of NFI member companies to comply with current law and regulation, and (2) developing an accountability system that would reward "good actors" and identify "bad actors." Such an accountability system involves screening by the Better Seafood Bureau, independent third-party audits of processes and products, and a member review process. The name was officially changed to the Better Seafood Board in May 2009 to avoid confusion with the Better Business Bureau.

As mentioned earlier in this report, the American Scallop Association, representing the majority of U.S. scallop production, has announced a plan to enforce labeling and packing standards for its 18 members to combat short-weight scallop packs.

In October 2012, more than 500 chefs and restaurant owners signed a petition, initiated by the environmental group Oceana, calling for the enactment of robust traceability standards for domestic and imported seafood to better protect consumers from illegal mislabeled fish.[52]

CURRENT LAW AND CONGRESSIONAL INTEREST

With the increasing media attention to this issue, Congress may face questions concerning current law applicable to seafood marketing and fraud. The issues to consider may include whether:

- current law applicable to fraudulent seafood sales and marketing is clear and enforceable;
- federal agency enforcement efforts targeting seafood fraud are adequate;
- penalties for seafood fraud offenses are a deterrent; and
- resources for federal agency enforcement are sufficient.

FDA is the primary agency responsible for ensuring that food sold in interstate commerce is properly labeled. The agency's jurisdiction includes seafood, and the agency operates an oversight compliance program, the Seafood Regulatory Program, for fishery products. Responsibility for a food product's safety, wholesomeness, identity, and economic integrity rests with the processor or importer, who must comply with regulations promulgated under the FFDCA and the FPLA. The agency has the authority to detain or temporarily hold food being imported into the United States while it determines if the product is misbranded or adulterated. The agency also has the authority to take legal action against sellers of adulterated and misbranded seafood and to recommend criminal prosecution or injunction of responsible firms and individuals. However, enforcement of economic fraud and labeling laws may be a lower FDA priority relative to protecting the health and safety of the U.S. food supply.[53]

The adequacy of agency funding is also an issue, especially whether the agency has adequate resources so that it can systematically monitor for economic fraud and mislabeling, determine the scope and scale of these types of problems, and develop new programs that address these concerns. About 85 of the agency's roughly 1,350 inspectors work primarily with seafood. The adequacy of agency funding is regularly raised as a concern when agency appropriations are considered. For example, in testimony before the Senate Committee on Appropriations, one industry representative stated:

> We believe the most important issue on the table today is whether FDA is adequately funded, has sufficient staff with scientific training and experience ... , has research dollars available to address key questions, has strong working agreements with the States to support as needed and has the commitment of the President and the Congress.[54]

Federal Food, Drug, and Cosmetic Act

The FFDCA attempts to keep interstate commerce free from misbranded (i.e., mislabeled) articles and to protect the public from inferior foods resembling standard products but marketed under distinctive names.[55] The governing statute for naming food is the FFDCA. The FFDCA, as amended, gives FDA authority over most food regulation and includes:

- a series of definitions elaborating on the concepts of adulteration and misbranding;
- control over all labeling of foods traveling in interstate commerce;
- detailed regulation of issues concerned with safety and wholesomeness of foods; and
- enforcement remedies available to the agency, when needed.[56]

An article is deemed misbranded if, among other things, its labeling is false or misleading or it is offered for sale under the name of another food.[57] An article is considered mislabeled when the label makes "no representation as to definition and standard of quality," unless the label bears the common or usual name of the food, if there is one.[58]

FDA has issued regulations that outline general principles for common or usual names of food. The common or usual name must:

> accurately identify or describe, in as simple and direct terms as possible, the basic nature of the food or its characterizing properties or ingredients. The name shall be uniform among all identical or similar products and may not be confusingly similar to the name of any other food that is not reasonably encompassed within the same name. Each class or subclass of food shall be given its own common or usual name that states, in clear terms, what it is in a way that distinguishes it from different foods.[59]

A common or usual name of a food may be established by common usage or by regulation.[60] For example, FDA promulgated a regulation establishing that Pacific whiting or North Pacific whiting is the common or usual name of the food fish *Merluccius productus*.[61] Most common or usual names, however, are established through common usage.

Although the agency's Seafood List includes "vernacular" names for some species, use of vernacular names is discouraged by the agency. The use of a vernacular name may cause a seafood to be misbranded under the FFDCA.[62] For example, the agency has issued specific guidance on using "red snapper" as a market name. The agency's policy states that "the labeling or sale of any fish other than *Lutjanus campechanus* as red snapper constitutes a misbranding in violation of the [FFDCA]."[63]

Under the FFDCA, species substitution also violates FDA's prohibition against adulteration. A food is deemed adulterated "if any substance has been substituted wholly or in part."[64] The marketing of a less valuable fish as one of higher value is a substitution and can result in a finding of adulteration.

Fair Packaging and Labeling Act

The FPLA requires that consumers of packaged commodities be provided with accurate information as to its contents. Congress passed the FPLA to "enable consumers to obtain accurate information as to the quantity of the contents and to facilitate value comparisons."[65] Under the FPLA, it is unlawful for persons engaged in labeling or packaging of consumer commodities "to distribute or to cause to be distributed in commerce any such commodity if such commodity is contained in a package, or if there is affixed to that commodity a label, which does not conform to the provisions" of the act.[66]

The FPLA requires each label to identify the commodity and the name of its manufacturer, packer, or distributor and the net quantity of contents, in terms of weight or mass, measure, or numerical count.[67] Food products falling within the scope of the FFDCA that are introduced into interstate commerce in violation of the FPLA and its regulations are deemed to be misbranded within the meaning of the FFDCA.[68]

State Regulation of Seafood Labeling

In addition to federal requirements, states also regulate the labeling and branding of seafood under state versions of the FFDCA. For example, in Alaska "no person may label or offer for sale any food fish product designated as halibut, with or without additional descriptive words, unless the food fish product is *Hippoglossus* or *Hippoglossus stenolepsis*."[69] In California, an individual who sells any commodity in less quantity than what is represented is guilty of a misdemeanor offense.[70]

Customs and Border Protection

Transshipment to avoid paying import or customs duties is illegal whenever it circumvents trade laws and other applicable trade restrictions.[71] The applicable law and regulation may vary, depending upon the trade agreement existing between the United States and another nation as well as the status of any antidumping and countervailing duties currently in force for particular products imported from designated nations.

In addition, it is a violation of the Lacey Act (16 U.S.C. 3372(d)) to falsely identify any fish, wildlife, or plant transported in interstate or foreign commerce.[72]

CONCLUSION

Both consumers and some producers could incur economic losses in cases of fraud or deception, with short-term gains to those who sell mislabeled merchandise. Although it is not clear whether the amount of fraud and deception in seafood sales and marketing is increasing, media attention to this issue has raised its profile with the public. The economic integrity initiative of the National Fisheries Institute has the potential to increase attention within the seafood industry to this issue as well as to address eroding consumer confidence in fair marketing of seafood produce. In addition, third-party certification bodies—for example, the Marine Stewardship Council,[73] Naturland,[74] and the Aquaculture Certification Council[75]—have the potential to address fraud and deception concerns where certification requires producers to adhere to specific protocols and standards for catching/raising and processing product.[76] However, the potential for these groups to self-enforce their standards has not been demonstrated.

In response to increased public concern, Congress is facing questions concerning current law applicable to seafood marketing and fraud. These questions include whether current law applicable to fraudulent seafood sales and marketing is clear and enforceable, whether federal agency enforcement efforts targeting seafood fraud are adequate, and whether the penalties for seafood fraud offenses are a deterrent. In addition, adequate funding may be an issue so that agencies can more systematically monitor the situation, better determine the scope and scale of this type of problem, and develop new programs that address this fraud.

In the 112th Congress, S. 50 proposed directing the Departments of Commerce and of Health and Human Services, the Federal Trade Commission, and other federal agencies to combat seafood fraud; on January 26, 2012, the Senate Committee on Commerce, Science, and Transportation reported this bill (S.Rept. 112-131), but no further action was taken. H.R. 6200 sought to address seafood fraud by requiring labels to identify species and origin for both domestic and imported fish and by requiring a plan to coordinate FDA and National Marine Fisheries Service seafood inspection.

GAO released a report on seafood fraud in February 2009, which may provide impetus for additional attention to FDA's role in addressing this issue.[77]

GAO concluded that federal agencies have not taken advantage of opportunities to share information that could benefit individual agency efforts to detect and prevent seafood fraud, and have not identified similar and sometimes overlapping activities that could be better coordinated to use limited resources more efficiently.

End Notes

[1] For a recent overview of this issue, see Bait and Switch: How Seafood Fraud Hurts Our Oceans, Our Wallets and Our Health, available at http://oceana.org/sites/default/ files/reports/ Bait_and_Switch_report_2011.pdf.

[2] Ian Doré, Seafood Scams and Frauds and How to Protect Yourself! (Toms River, NJ: Urner Barry Publications, Inc., 1992).

[3] Jann Th. Martinsohn, Deterring Illegal Activities in the Fisheries Sector: Genetics, Genomics, Chemistry and Forensics to Fight IUU Fishing and in Support of Fish Product Traceability, JRC Reference Report, Joint Research Centre, European Commission, 2011.

[4] In September 2008, NSIL was contacted by CRS, but was unable to provide more recent statistics.

[5] See http://sst.ifas.ufl.edu/22ndAnn/file08.pdf.

[6] Rosalee S. Rasmussen and Michael T. Morrissey, "DNA-Based Methods for the Identification of Commercial Fish and Seafood Species," Comprehensive Reviews in Food Science and Food Safety, v. 7 (2008): 280-295.

[7] For a technical description, see http://www.chemsoc.org/ExemplarChem/entries/2003/leeds_ chromatography/ chromatography/ief.htm; also see http://microbac.com/technical_articles/ news_detail.php?news_ID=14.

[8] See http://www.fishbol.org/index.php. The Fish Barcode of Life (FISH-BOL) Initiative is a global effort to assemble a standardized reference DNA sequence library for all fish species, and functions as a portal to the more comprehensive BoLD (Barcode of Life Database).

[9] See http://www.fda.gov/Food/FoodSafety/Product-SpecificInformation/Seafood/DNA specia tion/default.htm.

[10] For additional background on use of this technology, see Jann Th. Martinsohn, Deterring Illegal Activities in the Fisheries Sector: Genetics, Genomics, Chemistry and Forensics to Fight IUU Fishing and in Support of Fish Product Traceability, JRC Reference Report, Joint Research Centre, European Commission, 2011.

[11] "Wahoo Scam Yields Six Employees Sentenced to Probation," Samoa News (June 23, 2007); available at http://www.samoanews.com/monday/story3.html.

[12] For an Alaskan perspective on seafood fraud, see Barb Cooper, "Fish Faking," Alaska, v. 77, no. 10 (December 2011/January 2012): 32-42.

[13] For more information, see http://www.nmfs.noaa.gov/mediacenter/docs/ turbot_for_ halibut_fraud_may09.pdf.

[14] This is sometimes done to avoid U.S. import tariffs; for example, see http:// www. justice.gov/opa/pr/2008/October/ 08-enrd-967.html.

[15] For FDA import alert #16-04, see http://www.fda.gov/ora/fiars/ora_import_ia1604.html.

[16] For an example of mislabeling these imports, see http://agi.alabama.gov/press_releases/may-19-2006—sparks-warnsgulf-coast-businesses—restaurants-of-mislabeled-basa?pn=2. See also http://www.suffolkvanews.com/virginia/catfishconspiracy.php.

[17] P.B. Marko, et al., "Mislabelling of a Depleted Reef Fish," Nature, v. 430(2004):309-310. FDA policy states that labeling or sale of any fish other than Lutjanus campechanus as red snapper constitutes a misbranding in violation of the FFDCA.

[18] See http://wkrg.com/news/article/whats_on_your_plate/3561/ and http://wkrg.com/news/ article/is_it_really_grouper/ 3557/.

[19] See http://myfloridalegal.com/newsrel.nsf/newsreleases/C67 E4A15E6C1A3EA8525 74B 9004CF240.

[20] See http://pqasb.pqarchiver.com/sptimes/access/1230334761.html?dids=1230334761:1230334761&FMT=FT&FMTS=ABS:FT&date=Mar+10%2C+2007&author=CURTIS+KRUEGER&pub=St.+Petersburg+Times&edition=&startpage=1.A&desc=%27Grouper%27+costs+restaurants.

[21] Others have been reported in Portland, OR, http://www.kptv.com/print/11072970/detail.html; and in Phoenix, AZ, http://www.kpho.com/print/10371007/detail.html, for example.

[22] DNA barcoding is a relatively new technique that uses a short DNA sequence from a standardized and agreed-upon position in the genome as a molecular diagnostic for species-level identification. For additional information, see http://www.barcoding.si.edu/.

[23] John Schwartz, "Fish Tale Has DNA Hook: Students Find Bad Labels," The New York Times, August 22, 2008. Available at http://www.nytimes.com/2008/08/22/ science/ 22fish.html.

[24] For example, see American Fisheries Society, Common and Scientific Names of Fishes from the United States, Canada, and Mexico, Special Publication 29, 6th edition (Bethesda, MD: July 2004), 386 p.; and American Fisheries Society, World Fishes Important to North Americans, Special Publication 21 (Bethesda, MD: 1991), 243 p.

[25] See http://www.cfsan.fda.gov/~frf/seaintro.html.

[26] 63 Fed. Reg. 20,148 (April 23, 1998).

[27] Updated information on FDA concerns related to species substitutions and misbranding can be found at http://www.fda.gov/ora/fiars/ora_import_ia1604.html.

[28] See http://www.lawfuel.com/show-release.asp?ID=3593.

[29] Ami Petersen and David Green, Seafood Traceability: A Practical Guide for the U.S. Industry, North Carolina Sea Grant, Report UNC-SG-06-04, 28 p.; available at http://www.ncseagrant.org/files/seafood_traceability.pdf.

[30] 7 C.F.R. Part 60; See 69 Fed. Reg. 59708-59750 (October 5, 2004).

[31] 74 Fed. Reg. 2658-2707.

[32] See http://www.nola.com/news/?/base/news-2/1246253416178870.xml&coll=1.

[33] See http://www.noaanews.noaa.gov/stories2009/20090511_seafood.html.

[34] For more background on this practice, see James Wright, "The Weighting Game: The Seafood Industry Addresses Short-Weighting with Little Assistance," Seafood Business, July 2009, available at http://www.seafoodbusiness.com/ articledetail.aspx?id=4294994041.

[35] John Sackton, Seafood.Com News, August 12, 2008.

[36] See http://www.cfsan.fda.gov/~dms/seaglgui.html.

[37] For additional background on this investigation, see http://www.ncwm.net/sites/default/files/about/press/2010/ 2010_03_29_Seafood_Investigation.doc. For example, weights and measures inspectors of the Wisconsin Department of Agriculture, Trade and Consumer Protection found 157 short-weight violations at 23 grocery stores in seafood products from 10 companies, primarily from excessive ice glaze.

[38] See http://www.aboutseafood.com/press/press-releases/scallop-suppliers-embrace-economic-integrity.

[39] See http://www.fda.gov/bbs/topics/ANSWERS/ANS00422.html.

[40] See http://www.sushiman.net/ahi/carbon_monoxide_trea.htm.

[41] See CRS Report RL34247, Federal Regulation of Substances Generally Recognized As Safe (GRAS) and the Use of Carbon Monoxide in Packaging for Meat and Fish, by Vanessa K. Burrows and Cynthia Brougher.

[42] See FDA Import Bulletin #16B-95, available at http://seafood.ucdavis.edu/Guidelines/fdabulletin16b.htm.

[43] See http://www.cfsan.fda.gov/~rdb/opa-g015.html.

[44] Fereidoon Shahidi, Metusalach, and Joseph A. Brown, "Carotenoid Pigments in Seafoods and Aquaculture," Critical Reviews in Food Science and Nutrition, v. 28, no. 1 (1998): 1-67.

[45] 21 C.F.R. § 73.35; 21 C.F.R. § 73.75.

[46] A 2005-2006 study found that only 10 of 23 salmon fillets marketed as "wild" salmon were definitely caught in the wild, while the majority were from farmed salmon. See "Mislabeled Salmon: The Salmon Scam," Consumer Reports, v. 71, no. 8 (August 2006): 15.

[47] See http://www.atimes.com/atimes/Southeast_Asia/HC22Ae01.html and http://au.sys-con.com/node/890860/print. Further information is contained in out-of-print CRS Report RS21776, Shrimp Trade Dispute: Chronology, by Eugene H. Buck, available from the author.

[48] See http://www.noaanews.noaa.gov/stories2011/20110204_seafoodmislabeling.html.

[49] For further information, see CRS CRS Report RS21799, Saltonstall-Kennedy Act: A Summary of the Law and Its Implementation, by Eugene H. Buck.

[50] See http://www.southeasternfish.org/SPQC/product%20quality%20code.pdf.

[51] For more information, see http://www.aboutseafood.com/about/about-nfi/better-seafood-bureau, and http://www.aboutseafood.com/sites/www.aboutseafood.com/files/share/ BSB_Backgrounder_General_May_2009_NRA.pdf.

[52] See http://oceana.org/sites/default/files/reports/Oceana_Chef_Petition_0.pdf.

[53] "Species Substitution: Labeling Law Not An FDA Priority ... ," Santa Monica Seafood SeaLog (April 2006).

[54] Testimony by Thomas E. Stenzel, President and CEO of the United Fresh Produce Association, before the Senate Committee on Appropriations, March 12, 2007, in a special hearing on "Food Safety.".

[55] 35A Am. Jur. 2d Food § 25 (2006).

[56] For more information, see CRS Report RS22600, The Federal Food Safety System: A Primer, by Renée Johnson.

[57] 21 U.S.C. § 343. FDA further regulates this area of "Misbranding" in 21 C.F.R. § 101.18.

[58] 21 U.S.C. § 343(i)(1).

[59] 21 C.F.R. § 102.5(a).

[60] 21 C.F.R. § 102.5(d).

[61] 21 C.F.R. § 102.46.

[62] Sec. 540.750 Common or Usual Names for Seafood in Interstate Commerce (CPG 7108.26).

[63] Sec. 540.475 Snapper—Labeling (CPG 7108.21).

[64] 21 U.S.C. § 342 (b)(2).

[65] 15 U.S.C. § 1451.

[66] 15 U.S.C. § 1452.

[67] 15 U.S.C. § 1453(a).

[68] 15 U.S.C. § 1456(a).

[69] Alaska Stat. § 17.20.045.

[70] Cal Bus & Prof Code § 12024.

[71] Section 592 of the Tariff Act of 1930 (19 U.S.C. § 1592).

[72] For an example of a case prosecuted under this provision, see http://www.justice.gov/usao/waw/press/2011/apr/ jay.html.

[73] See http://www.msc.org/about-us.

[74] See http://www.naturland.de/certification_aquaculture.html.

[75] See http://www.aquaculturecertification.org/.

[76] Problems with supply chain integrity that potentially can compromise such certification programs have been reported, see Peter B. Marko, Holly A. Nance, and Kimberly D.

Guynn, "Genetic Detection of Mislabeled Fish from a Certified Sustainable Fishery," Current Biology, v. 21, no. 16 (August 23, 2011): R621-R622.

[77] U.S. Government Accountability Office, Seafood Fraud: FDA Program Changes and Better Collaboration among Key Federal Agencies Could Improve Detection and Prevention, February 2009, GAO-09-258, 49 p.

In: Food Fraud and Adulterated Ingredients
Editor: Darrell T. Braden
ISBN: 978-1-63117-730-9

Chapter 3

FOOD AND DRUG ADMINISTRATION: BETTER COORDINATION COULD ENHANCE EFFORTS TO ADDRESS ECONOMIC ADULTERATION AND PROTECT THE PUBLIC HEALTH

United States Government Accountability Office

WHY GAO DID THIS STUDY

In recent years, the United States experienced public health crises suspected to have been caused by the deliberate substitution or addition of harmful ingredients in food and drugs—specifically melamine in pet food and oversulfated chondroitin sulfate in the blood thinner heparin. These ingredients were evidently added to increase the apparent value of these products or reduce their production costs, an activity GAO refers to as economic adulteration. The Food and Drug Administration (FDA), an agency within the Department of Health and Human Services (HHS), has responsibility for protecting public health by ensuring the safety of a wide range of products that are vulnerable to economic adulteration. This report examines (1) the approaches that FDA uses to detect and prevent economic adulteration of food and medical products and (2) the challenges FDA faces in detecting and preventing economic adulteration and views of stakeholders on options for FDA to enhance its efforts to address economic adulteration. GAO reviewed

FDA documents and interviewed FDA officials and stakeholders from academia and industry, among others.

WHAT GAO RECOMMENDS

GAO recommends that FDA adopt a working definition of economic adulteration, enhance communication and coordination of agency efforts, and provide guidance to agency centers and offices on the means of addressing economic adulteration. HHS neither agreed nor disagreed with GAO's recommendations, but cited planned actions related to adopting a definition and enhancing communication and coordination.

WHAT GAO FOUND

FDA primarily approaches economic adulteration as part of its broader efforts to combat adulteration in general, such as efforts to ensure the safety of imported products. Agency officials noted that the Federal Food, Drug, and Cosmetic Act does not distinguish among motives or require motive to be established to determine whether a product is adulterated. However, a senior FDA official told GAO that there is value in making a distinction between economic adulteration and other forms of adulteration to guide the agency's thinking about how to be more proactive in addressing this issue. An FDA official told GAO when the agency detects any form of adulteration that poses an adverse public health effect, it can conduct an investigation, request a recall to get the product off the market, and take enforcement action. In addition to these broader efforts, some FDA entities also have undertaken efforts that specifically focus on economic adulteration. For example, FDA's Office of Regulatory Affairs has contracted with a research center to model risk factors for improved detection of economic adulteration of food. However, FDA entities have not always communicated or coordinated their economic adulteration efforts. For example, FDA's Center for Veterinary Medicine was unaware of and did not participate in two other entities' economic adulteration efforts involving products the veterinary center regulates. In another instance, two FDA entities engaged in similar efforts but did not communicate or coordinate them, even though officials said such communication might be beneficial. Furthermore, FDA has not issued specific written guidance on how its centers and offices should approach or address their economic adulteration efforts. This is not consistent with federal standards for internal control, which require agencies to have documented policies and procedures.

FDA officials and stakeholders GAO interviewed cited several key challenges to detecting and preventing economic adulteration, including increased globalization and lack of information from industry. Globalization has led to an increase in the variety, complexity, and volume of imported food and drugs, which complicates FDA's task of ensuring their safety. In addition to globalization, an increase in supply chain complexity—the growth in the networks of handlers, suppliers, and middlemen—also complicates FDA's task, making it difficult to trace an ingredient back to its source. FDA officials and stakeholders also said that gathering information from industry, such as information on potentially adulterated ingredients, presents challenges for FDA in detecting and preventing economic adulteration due to industry's reluctance to share such information because it is proprietary. Stakeholders cited greater oversight and information sharing as options to improve FDA's ability to combat economic adulteration. Specifically, some stakeholders supported increased oversight, such as the use of technology to trace adulterated ingredients back to the point of contamination, as an option to obtain more information on supply chains. Many stakeholders also suggested that FDA increase its regulatory and enforcement actions to address economic adulteration, including in instances that may not have a large negative public health impact. Stakeholders also suggested that greater communication with industry, through such means as an information clearinghouse or more informal interactions, could enhance FDA efforts to gather information on economic adulteration.

ABBREVIATIONS

API	active pharmaceutical ingredient
CBER	Center for Biologics Evaluation and Research
CDER	Center for Drug Evaluation and Research
CDRH	Center for Devices and Radiological Health
CFSAN	Center for Food Safety and Applied Nutrition
CVM	Center for Veterinary Medicine
FDA	Food and Drug Administration
HHS	Department of Health and Human Services
ORA	Office of Regulatory Affairs

PREDICT Predictive Risk-based Evaluation for Dynamic Import Compliance Targeting
USP United States Pharmacopeia

October 24, 2011

The Honorable Henry A. Waxman
Ranking Member
Committee on Energy and Commerce
House of Representatives

The Honorable Frank Pallone, Jr.
Ranking Member
Subcommittee on Health
Committee on Energy and Commerce
House of Representatives

The Honorable John D. Dingell
House of Representatives

In recent years, the United States experienced two crises suspected to have been caused by the deliberate substitution or addition of harmful ingredients in food and drugs. These ingredients were evidently added to increase the apparent value of these products or reduce their production costs, an activity we refer to as "economic adulteration." Specifically, in 2007, vegetable protein products imported to the United States from China were found to contain melamine, an industrial chemical, which later investigation suggested had been added to give the appearance of higher protein content. The products were used as ingredients in pet food, sickening and killing an unknown number of dogs and cats.[1] In 2008, heparin—a commonly used blood thinner—was imported from China and found to contain oversulfated chondroitin sulfate, a toxic contaminant that mimics heparin. The contaminated heparin was linked to a number of serious allergic reactions and deaths in the United States. In addition to harming public health, such incidents can undermine confidence in the safety of the nation's food and medical products and have significant economic consequences for industry. For example, the melamine incident prompted pet food manufacturers to recall over 150 brands of dog and cat food across the United States, and the heparin crisis prompted 16 drug and device firms to recall at least 11 drug products and 72 heparin-containing medical devices.

The Food and Drug Administration (FDA), an agency within the Department of Health and Human Services (HHS), has responsibility for protecting public health by ensuring the safety of a wide range of food and medical products (drugs, medical devices, and biologics[2]) that are vulnerable to, among other things, economic adulteration. With increasing globalization of food production and medical product manufacturing, the volume of imported goods regulated by FDA has more than doubled in the last decade. As a result, according to FDA documents, from 10 to 15 percent of all food consumed in the United States is now imported, as are about 80 percent of the active pharmaceutical ingredients in drugs, nearly 40 percent of finished drugs, and half of all medical devices. FDA expects the growth in imports to continue. The FDA Commissioner has said that globalization presents huge and growing challenges and that economic adulteration remains a public health threat. She indicated that another public health crisis like heparin or melamine seems inevitable unless FDA is able to forge changes in how it ensures the safety and quality of food and medical products. In part due to regulatory challenges posed by increased globalization, we included federal oversight of both food and medical products in our 2011 list of federal programs at high risk for waste, fraud, abuse, or mismanagement that warrant attention by Congress and the executive branch.[3] We first added federal oversight of food safety to our High-Risk list in January 2007 and federal oversight of medical products in January 2009.[4]

This report responds to your request that we review how FDA oversees the safety of food and drugs in order to prevent and respond to economic adulteration. This report examines (1) the approaches that FDA uses to detect and prevent economic adulteration of food and medical products and (2) the challenges, if any, FDA faces in detecting and preventing economic adulteration and stakeholder views on options for FDA to enhance its efforts to address economic adulteration.

For purposes of this report, we define economic adulteration as "the fraudulent, intentional substitution or addition of a substance in a product for the purpose of increasing the apparent value of the product or reducing the cost of its production, i.e., for economic gain. [It] includes dilution of products with increased quantities of an already-present substance (e.g., increasing inactive ingredients of a drug with a resulting reduction in strength of the finished product, or watering down of juice) to the extent that such dilution poses a known or possible health risk to consumers, as well as the addition or substitution of substances in order to mask dilution." This definition is the same as the working definition of "economically motivated adulteration" that FDA developed for a May 2009 public meeting[5] to raise awareness and solicit input on the topic.[6]

To determine the approaches FDA uses to detect and prevent economic adulteration, we interviewed FDA officials about the agency's efforts to address economic adulteration and reviewed relevant FDA documents. We also reviewed our previous reports on FDA's oversight of food and medical products, as well as the agency's strategic planning efforts. We compared FDA's efforts to address economic adulteration with federal standards for internal control. To determine the challenges FDA faces in detecting and preventing economic adulteration, we interviewed FDA officials and stakeholders, including former FDA officials and some representatives from academia, industry, and consumer groups who made presentations at FDA's May 2009 meeting on economically motivated adulteration. We also interviewed the stakeholders to obtain their views on options for FDA to enhance its efforts to address economic adulteration. The views of these stakeholders are not representative of and cannot be generalized to all stakeholders. Appendix I contains a detailed discussion of the scope and methodology of our review.

We conducted this performance audit from September 2010 to October 2011 in accordance with generally accepted government auditing standards. Those standards require that we plan and perform the audit to obtain sufficient, appropriate evidence to provide a reasonable basis for our findings and conclusions based on our audit objectives. We believe that the evidence obtained provides a reasonable basis for our findings and conclusions based on our audit objectives.

BACKGROUND

FDA is responsible for protecting public health by ensuring the safety and efficacy of medical products marketed in the United States—including drugs, medical devices, and biologics—and the safety of nearly all food products other than meat and poultry,[7] regardless of whether they were manufactured domestically or overseas. The agency's responsibilities for overseeing food and medical products are divided among the following five FDA product centers, each responsible for specific types of products:

- The Center for Biologics Evaluation and Research (CBER) is responsible for regulating biologics for human use, such as blood, blood products, vaccines, and allergenic products, and ensuring that biologics are safe and effective.
- The Center for Devices and Radiological Health (CDRH) is responsible for regulating firms that manufacture and import medical devices and for ensuring that radiation-emitting products, such as lasers and x-ray systems, meet radiation safety standards.
- The Center for Drug Evaluation and Research (CDER) is responsible for regulating over-the-counter and prescription drugs for human use, including generic drugs.
- The Center for Food Safety and Applied Nutrition (CFSAN) is responsible for ensuring the safety of most foods for humans (except meat and poultry and processed egg products, which are regulated by the U.S. Department of Agriculture), including dietary supplements.
- The Center for Veterinary Medicine (CVM) is responsible for regulating the manufacture and distribution of drugs, devices, and food given to, or used by animals.

Among other things, the centers monitor the safety and effectiveness of marketed medical products and the safety of food, formulate regulations and guidance, conduct research, communicate information to industry and the public, and set their respective program priorities.

In addition to the work of the five centers, FDA's Office of Regulatory Affairs (ORA) conducts field work for the product centers to promote compliance with agency requirements and applicable laws. ORA field activities include inspecting domestic and foreign manufacturing facilities, examining products offered for import, collecting and analyzing samples, and taking enforcement action.[8] ORA's Office of Criminal Investigations is responsible for investigating potential criminal violations involving FDA-regulated products and may refer cases to the Department of Justice for prosecution.

FDA's Office of the Commissioner is responsible for providing leadership and direction to the product centers and ORA. FDA's Office of International Programs is responsible for leading, managing, and coordinating all of FDA's international activities and its recently established overseas offices.

In July 2011, FDA created "directorates" that align similar functions under common leadership within the Office of the Commissioner—the Office of Medical Products and Tobacco, which oversees CBER, CDER, and CDRH, as well as the Center for Tobacco Products; the previously established Office of Foods, which oversees CFSAN and CVM; and the Office of Global Regulatory Operations and Policy, which oversees ORA and the Office of International Programs.

In recent years, we have reported on a variety of concerns related to FDA's resource management, strategic planning, and internal communications and coordination. Specifically, in June 2009, we found that FDA was unable to provide complete and reliable estimates of its resource needs for its medical products.[9] In February 2010, we reported on management challenges the agency faces and FDA's difficulties in using practices for effective strategic and workforce planning.[10] Coordinating internally among its centers and offices and externally with outside experts were among the agency's major management challenges. Also, in September 2010, we reported on FDA's overseas offices and the need for better coordination among the centers.[11] For a list of these and other related reports, see Related GAO Products at the end of this report.

Federal Law on the Adulteration of FDA- Regulated Food and Medical Products

The Federal Food, Drug, and Cosmetic Act prohibits the introduction of adulterated food, drugs, and medical devices into interstate commerce. However, the act does not define or use the term "economic adulteration" or "economically motivated adulteration." The act includes, but is not limited to, the following:

- A food is deemed to be adulterated if, among other circumstances, it bears or contains any added poisonous or deleterious substance that may render it injurious to health. A food is also deemed to be adulterated (1) if any valuable constituent has been omitted in whole or in part, or (2) if any substance has been substituted wholly or in part, or (3) if damage or inferiority has been concealed in any manner, or (4) if any substance has been added so as to increase its bulk or weight, or reduce its quality or strength, or make it appear better or of greater value than it is.
- A drug is deemed to be adulterated if it purports to be a drug whose name is recognized in an official compendium[12] and its strength differs from, or its quality or purity falls below, the standards set forth

in such compendium. If a drug does not purport to be a drug listed in an official compendium, it is deemed to be adulterated if its strength differs from, or its purity or quality falls below, that which it purports to possess. A drug is also deemed to be adulterated if, among other circumstances, any substance has been (1) mixed or packed with it so as to reduce its quality or strength or (2) substituted wholly or in part.

- A device is deemed to be adulterated if it is, or purports to be or is represented as, a device which is subject to a performance standard established or recognized under the act unless such device is in all respects in conformity with such standard. It is also deemed adulterated if, among other circumstances, the device was not manufactured, packed, stored, or installed in conformity with good manufacturing practices.

Economic Adulteration

Economic adulteration is not a new problem and ranges from simple actions, such as adding material to increase a product's weight, to more sophisticated substitutions or additions that are designed to avoid detection by tests known to be used to authenticate ingredients or products. Economic adulteration differs from other forms of intentional adulteration, such as bioterrorism or sabotage, whose primary purpose is to cause harm. Because economic adulteration is intentional, it also differs from unintentional adulteration, such as adulteration through failure to follow good manufacturing practices.

Although the primary driver of economic adulteration is financial gain rather than causing harm, it can pose a variety of public health risks. The actual risks will vary depending on the adulterant used, the person who consumes the product, and the length of use or exposure. There is a direct and immediate threat to public health when the adulterant is a toxic or lethal substance, as was the case in the melamine and heparin incidents. There are also risks that arise as a result of long-term, low-dosage exposure to a contaminant or as a result of using a product whose nutritional value or efficacy has been compromised by an adulterant. Certain populations, such as infants, the elderly, and persons with compromised immune systems are particularly vulnerable to these risks. In some cases, an adulterant may only pose a public health risk for those who are allergic to it, such as fish substituted with a less expensive fish to which a person is allergic.

Furthermore, economic adulteration that poses no known health risk may expose a vulnerability in the supply chain—the network of handlers, suppliers, and middlemen involved in the production of food and drugs—that could be further exploited in the future, with serious consequences.

Following the melamine and heparin incidents, FDA formed an internal work group focused on predicting and addressing what the agency referred to as "economically motivated adulteration." The work group, comprising representatives from FDA's food and medical product centers and ORA, held a May 2009 public meeting on the topic. For purposes of the meeting, FDA developed a working definition of economically motivated adulteration. The meeting, attended by representatives of academia, industry, and consumer groups, was designed to raise awareness about the potential for this problem and gather information on how to better predict, prevent, and address it. According to FDA officials, the work group stopped meeting shortly after the public meeting was held. FDA made a transcript of the meeting publicly available, but issued no report.

FDA Has Many Efforts to Address Economic Adulteration but Has Missed Opportunities to Communicate and Coordinate

FDA primarily approaches economic adulteration as part of its broader efforts to detect and prevent adulteration of food and medical products in general. In addition, CDER, ORA, CFSAN, and CBER have undertaken efforts specific to economic adulteration, while CVM and CDRH have not. However, agency entities have missed opportunities to communicate and coordinate efforts specifically directed at economic adulteration and identify potential public health risks.

FDA Primarily Approaches Adulteration Broadly, but Some Centers and ORA Have Undertaken Efforts Specific to Economic Adulteration

According to FDA officials, the agency primarily approaches economic adulteration as part of its broader efforts to combat adulteration in general. Such efforts include, for example, the agency's actions to ensure the safety of imported products. According to FDA officials, these broad efforts to combat adulteration could also combat economic adulteration. Agency officials noted that the Federal Food, Drug, and Cosmetic Act does not distinguish among motives or require motive to be established to determine whether a product is adulterated. FDA adopted a working definition of economically motivated adulteration for the purposes of discussing the topic at its May 2009 public meeting. In its written comments on our draft report, HHS told us that the recently formed FDA Working Group on Economically Motivated Adulteration will use the working definition proposed at the public meeting, enabling FDA centers to focus their discussions and encouraging communication and collaboration. According to an FDA official, the agency generally does not expend resources to distinguish between economic and other motives for adulteration. Rather, when the agency detects any form of adulteration that poses an adverse public health effect, it can conduct an investigation, request a recall to get the product off the market, and take enforcement action. A senior FDA official told us there is value in making a distinction between economic adulteration and other forms of adulteration to guide the agency's thinking about how to be more proactive in addressing this issue.

Examples of broader FDA efforts to address adulteration include:

- *ORA's Predictive Risk-Based Evaluation for Dynamic Import Compliance Targeting (PREDICT).* This tool generates a numerical risk score for all FDA-regulated products by analyzing importers' shipment information using sets of FDA-developed risk criteria based in part on publicly available information,[13] which may indicate opportunities for economic adulteration. PREDICT then targets for examination products that have a high risk score. As of September 2011, PREDICT was operating in ports of entry in 13 of 16 FDA districts, and FDA officials said the agency expects PREDICT to be operational in all ports of entry by the end of 2011.
- *CVM's Pet Event Tracking Network (PETNet).* In August 2011, CVM launched PETNet, a secure, Internet-based network comprised of FDA and other federal and state agencies with authority over pet food that would allow them to exchange real-time information about outbreaks of illness in animals associated with pet food and other pet food-related incidents. PETNet members can elect to receive alerts about pet food incidents and create alerts when they are aware of a pet food incident within their jurisdiction. According to FDA, the

information would be used to help federal and state regulators determine how best to use inspectional and other resources to either prevent or quickly limit the adverse events caused by adulterated pet food. Use of the system is voluntary.

- *CDER's Secure Supply Chain Pilot Program.* This program, which is in the process of being implemented, is intended to help the agency ensure the safety of imported drugs by enabling it to focus its resources on preventing the importation of drugs that do not comply with applicable FDA requirements. The program is intended to allow a limited number of drug companies to import their products on an expedited basis if, among other things, they can meet FDA criteria showing that they maintain control over their products from manufacture through entry into the United States. FDA expects to announce the date on which it will begin accepting applications for the pilot by the end of 2011.

In addition to these broader efforts, some FDA entities have undertaken efforts specific to economic adulteration. For example, in the aftermath of the melamine and heparin incidents, CDER, ORA, CFSAN, and CBER have taken the following steps to specifically address economic adulteration:

- CDER has developed a model to rank the 1,387 active pharmaceutical ingredients (API) known to be in current use according to their susceptibility to economic adulteration.[14] According to CDER officials, the ranking model incorporates various risk factors, such as estimates for volume of use, cost per unit of the API, and reliance on testing methods to check quality that are known to be less accurate than more modern methods developed for other APIs. CDER officials told us the center sampled and tested 20 of the 77 higher- ranked APIs in 2010 and found no evidence of any significant contamination suggesting intentional adulteration. According to agency officials, after this pilot program is completed, FDA will determine if the program was valuable and, if so, whether the model's risk factors may need to be adjusted.
- CDER is leading efforts to work with United States Pharmacopeia (USP) to focus on the vulnerability of drugs to economic adulteration. USP is a nonprofit organization that sets standards for medicines, food ingredients, and dietary supplements. USP's drug standards are enforceable under the Federal Food, Drug, and Cosmetic Act. Actions

CDER officials say they have taken include selecting 20 USP standards[15] for updating that include certain over-the-counter drugs, inactive ingredients used in high volume, and APIs that use outdated technology or for which there are no procedures to identify impurities. The goal of this modernization effort is to replace outdated USP standards with more modern, accurate, and specific ones. CDER has also worked closely with USP in revising the heparin testing standard and the standards for glycerin and five other similar drug product ingredients to prevent economic adulteration with diethylene glycol, a cheaper, but deadly ingredient often substituted for glycerin.

- ORA, along with the Department of Homeland Security and the Department of Agriculture's Food Safety and Inspection Service, contracted in 2010 with the University of Minnesota's National Center for Food Protection and Defense to model risk factors for improved detection of economic adulteration.[16] The contract consists of three phases: (1) a survey of U.S. companies to collect information on prior or potential economic adulteration experiences and identify characteristics of potential targets of economic adulteration; (2) the development of strategies to group test methods to identify those methods that pose the greatest potential risk for economic adulteration, including the level of technical sophistication required to exploit the test method; and (3) the development of supply chain models in order to identify shifts in these supply chains that may indicate the potential for economic adulteration.
- CFSAN formed a work group on the economic adulteration of food, which started meeting in February 2008. CFSAN officials said the group, which includes representatives from CVM and ORA, generally meets monthly and is looking at the impact of economic adulteration on food safety and whether there is other work that FDA could undertake to mitigate that impact. Among other things, the group has proposed creating a page on FDA's website on economic adulteration, and it has developed a methodology for the testing of pomegranate juice, which officials said they chose to focus on because it is expensive and because its health benefits have been widely touted. CFSAN officials said the group is also looking at ways to make industry more comfortable with providing information to FDA on possible economic adulteration.
- CFSAN has a number of efforts under way to develop analytical tests and tools for detecting economic adulteration. For example, the center

has developed a method for analyzing nitrogen-containing compounds similar to melamine that might be used to boost apparent nitrogen content in milk and other protein products. CFSAN's research office has a project to develop methods to detect the adulteration of powdered milk products and fruit juices. One element of this project involves the creation of a library of powdered milk signatures, against which new samples can be compared (and adulterants identified) using modern statistical methods. This project is slated for completion over the next 2 years.

- CBER established a process in late 2008 to extract relevant product component information from regulatory applications and input this data into a database. The center has since expanded its process of extracting product component information from applications to include, for example, ingredients that may be subject to contamination. The database includes nonproprietary, unique ingredient identifiers and other information designed to facilitate faster identification of products made from components suspected of being economically adulterated.

In contrast to the other entities, senior CVM officials we spoke with said that although the center has broad initiatives designed to prevent and detect adulteration in general (i.e., PETNet), CVM has undertaken no efforts targeted to economic adulteration and has no plans to do so.[17] Officials said that the melamine incident gave them greater awareness that products with high-value ingredients could be susceptible to economic adulteration and that this was the only lesson they learned from the melamine incident. Officials said they recognize that CVM-regulated products may be vulnerable to economic adulteration because they are composed of numerous byproducts, any one of which could be adulterated. Nevertheless, they said that they do not believe economic adulteration is a growing problem because of industry's overall awareness of its supply chain through efforts such as verifying certificates of analysis of ingredients from suppliers.

Officials from CDRH told us that, other than its broader efforts to combat adulteration in general, the center had no initiatives specifically directed at addressing economic adulteration, but indicated they are responsible for products that are vulnerable to economic adulteration. For example, CDRH officials said that they have found that a manufacturer of imported sunglass lenses may have been substituting inferior material. However, center officials were unaware of any actual cases of economic adulteration involving products for which they are responsible.

Agency Entities Have Not Communicated or Coordinated on Economic Adulteration

We found two instances in which CVM did not know about or participate in efforts on economic adulteration that involved CVM-regulated products. First, the director of the University of Minnesota's National Center for Food Protection and Defense told us that, as part of the center's contract with ORA, it will be drafting a list of foods at high risk of economic adulteration and that the list will likely include foods that are also used as animal feed ingredients. The director noted that, with the exception of certain kinds of fats, the global supply chain for animal food and feed is the same as that for human foods. The director said that, for this reason, his center had considered finding ways to make its work for FDA even more applicable to animal feed. Although CVM provided developmental input, direction, and technical support with regard to the contract, CVM officials said they were not aware of the center's work under the contract to develop this list of high-risk foods. Second, CFSAN has a research project that focuses primarily on developing methods for authenticating protein-based foods and ingredients, detecting the presence of adulterants, and identifying chemical hazards in protein-based products. Among other things, this project is to develop methods for screening skim milk powder, which can be found in both food and animal feed, for the presence of soy or other vegetable protein. Senior CVM officials said they were unaware of this research project, but they stated that CVM has been involved in developing methods to identify contaminants of protein-based ingredients.

We also found an instance where FDA entities engaged in similar efforts on economic adulteration but did not communicate or coordinate about those efforts. Specifically, as we mentioned earlier, ORA and CDER are engaged in similar efforts to determine which human foods and drugs, respectively, are at greatest risk for economic adulteration. However, according to ORA and CDER officials, they have not coordinated those efforts or communicated about them, even though they are using some of the same risk factors in their efforts—including price fluctuations and reliance on less specific test methods. Officials from both entities said that such communication and coordination could be beneficial to both efforts.

In addition, we have previously identified internal coordination—among FDA's centers and offices—as one of the agency's major management challenges based on a review of evaluations of FDA by HHS and the FDA Science Board, among others.[18] Also, in our 2009 survey of FDA managers, 70 percent reported that better internal coordination and communication would greatly improve their ability to contribute to FDA's goals and responsibilities, though 28 percent reported that FDA was making great progress in this area. Furthermore, we asked FDA managers in our survey to identify the top priorities that FDA leadership should address to achieve agency goals and responsibilities, and the second most commonly identified issue was improving coordination within FDA. In detailed written responses in our survey, some managers noted that better coordination among FDA's centers could increase effectiveness and decrease redundancy.

Furthermore, a recommendation made by FDA's work group on economic adulteration in August 2009 related to communication—that FDA designate a lead office and develop standard operating procedures for information sharing—was not implemented. A senior FDA official told us that there has been some work across FDA centers on economic adulteration but that the centers did not see a lot of value in additional coordination because of the differences between the products each center oversees. However, the issue of economic adulteration cuts across the agency, and without communicating about and coordinating on economic adulteration efforts, FDA may not be making the best use of scarce resources.

In August 2011, FDA officials told us that the agency's Compliance Policy Council, which consists of senior representatives of ORA and the FDA centers, met in July 2011 and discussed whether and how the agency should coordinate work on economic adulteration. The council directed risk management staff from ORA and the centers to form a group to discuss opportunities to share intelligence and approaches to economic adulteration and then report back to the council. According to FDA officials, the proposed agenda included discussion about the development of standard operating procedures. In its written comments on our draft report, HHS told us that the work group held its first meeting on September 23, 2011, while our report was at the agency for comment.

The Commissioner and other senior FDA officials have often spoken publicly about the threat posed by economic adulteration. In its July 2011 report entitled *Pathway to Global Product Safety and Quality*,[19] FDA stated that globalization has fundamentally altered the economic and security landscape, requiring FDA to transform itself into a global agency prepared to regulate in an environment in which product safety and quality know no borders. The report also called economically motivated harms perhaps the most serious challenge on the horizon for the agency and noted that the heparin and melamine incidents underscore how serious the potential danger can be. The report also noted that FDA needs to move beyond its current efforts and think strategically across the agency. However, FDA officials told us that the Office of the Commissioner has not issued specific written guidance on how FDA centers and offices should approach or address their economic adulteration efforts. The Office of the Commissioner's role is to provide policy making, program direction, coordination, liaison, and expert advice for agency programs. According to federal standards for internal control, agencies should have documented policies and procedures in place to carry out management's directives.[20] This documentation should be readily available for examination in management directives, administrative policies, or operating manuals in paper or electronic form. In addition, the federal standards call for effective communication, with information flowing down, across, and up the organization.

FDA Faces Challenges in Addressing Economic Adulteration, and Stakeholders Identified Options That May Help Agency Efforts

FDA officials and stakeholders we interviewed cited several key challenges the agency faces in detecting and preventing economic adulteration, and stakeholders identified options for enhancing the agency's efforts to address economic adulteration.

Key Challenges Include Increased Globalization and Lack of Information from Industry

FDA officials and stakeholders told us that responding to increased globalization and the expanding complexity of the supply chains for both food and medical products is a key challenge in addressing economic adulteration. Globalization has led to an increase in the variety, complexity, and volume of imported food and drugs, which complicates FDA's task of ensuring their safety. In addition to globalization, an increase in supply chain complexity—

the growth in the networks of handlers, suppliers, and middlemen—also complicates FDA's task. According to FDA, the market for outsourcing portions of pharmaceutical production has more than doubled in the past 9 years. FDA noted in its July 2011 *Pathway* report that more products are following increasingly complex paths through multi-step supply chains before reaching the United States. Figure 1 illustrates the complex supply chain of a single commodity, canned tuna. As the figure shows, after the tuna is caught in East Asia, it can travel through many countries for processing and canning before the finished product finally reaches store shelves in the United States.

FDA officials gave several reasons that this increasing complexity poses a challenge. For example, CFSAN officials told us that food companies can change ingredients and suppliers at will without having to notify FDA of those changes, making it difficult to track or trace an ingredient back to its source or supplier. However, many food manufacturers are required to keep records of the immediate previous sources of all foods received. Similarly, CDER officials said that it is increasingly difficult to trace ingredients through drug supply chains due to the increasing number of parties involved and the increase in transfers between parties in other countries. Stakeholders from associations representing the food and medical product industries agreed that the large number of imported ingredients and foreign establishments, as well as the difficulties related to tracking an ingredient back to the original source, are of particular concern.

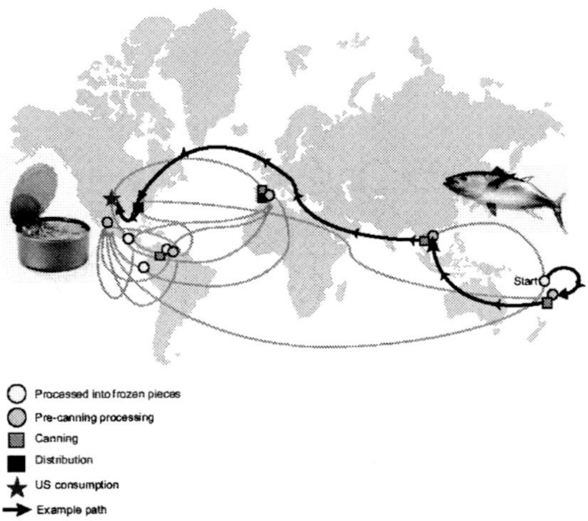

Source: FDA.

Figure 1. Illustrative Supply Chain for Canned Tuna.

FDA officials and stakeholders said that obtaining information on potential instances of economic adulteration is critical to addressing the problem, but they also agreed that the agency faces challenges in gathering such information from industry. Industry may be a source of information on potential incidents of adulteration because companies regularly test ingredients from suppliers. The responsible party for a firm that introduces into commerce an article of food containing an adulterated ingredient that could cause serious adverse health consequences or death must report this information to FDA through the Reportable Food Registry. However, agency officials and industry representatives said industry is often reluctant to share such information when an adulterated ingredient has not entered into commerce. For example, a company may be concerned that it could provoke a lawsuit if it reported a supplier for intentionally adulterating products and the accusation was subsequently determined to be unfounded. They said that a wrongful accusation can have serious consequences, such as compromising the integrity of the company's brands and products if certain information became public.

In addition to a need for more information about industry suppliers, FDA officials told us that they need more information about substances that could be used to adulterate products. These officials said that new, more precise testing methods need to be developed to detect these adulterants because some current tests are outdated or insufficiently specific. Recent cases of melamine contamination in pet food illustrate the need for such tests. The presence of melamine in pet food was not initially discovered by the standard test for protein because that test was designed to detect nitrogen and could not distinguish between protein and melamine. The contamination was ultimately discovered when FDA scientists developed a specific test to identify melamine. FDA and others determined that melamine was apparently selected as an adulterant to evade the original testing and increase the apparent protein content.

CDER officials also told us that it is difficult to detect instances of economic adulteration because the potential adulterant is often unknown or has not yet been identified. For example, during the heparin incident, the available test methods for heparin were not able to detect the contaminant oversulfated chond roitin sulfate. FDA collaborated with scientists outside the agency to identify the contaminant and develop new tests to detect it. Industry may be the best source of tests to detect adulteration because companies develop such tests to monitor the products they receive from their suppliers; however, industry officials indicated that they are often reluctant to share such information because it is proprietary.

Stakeholders cited additional challenges that FDA faces in addressing economic adulteration, including its legal authorities. For example, one stakeholder said that FDA does not have the authority to accredit, or approve third parties to inspect establishments that make drugs; the stakeholder said that if FDA did have that authority, such inspections may help decrease FDA's inspection workload and could increase the total number of facilities inspected. FDA recently received authority to recognize, in certain situations, accreditation bodies that may then accredit qualified third parties to inspect food establishments. The FDA Food Safety Modernization Act provides that, no later than January 2013, FDA is to establish a program to recognize these accreditation bodies. It is worth noting, though, that FDA has had the authority to accredit third parties to conduct inspections of certain domestic and foreign medical device manufacturing establishments since 2002. FDA implemented its accreditation programs, permitting eligible establishments to voluntarily request inspections from third-party organizations, but relatively few establishments have chosen to take advantage of this program.[21]

Some stakeholders also told us that FDA's limited resources, including staffing, present a challenge. Specifically, they said FDA has limited ability to investigate potentially economically adulterated products because such investigations are resource-intensive. They also told us that FDA does not have the range of expertise among staff that is needed to address economic adulteration, in particular staff with a background in intelligence gathering or law enforcement. We have previously reported on FDA's own concerns about its staffing levels and oversight responsibilities for certain activities, such as its oversight of medical devices and inspections of establishments that manufacture approved drugs.[22]

Some Stakeholders Cited Greater Oversight and Information Sharing as Options to Address Economic

Some stakeholders supported increased oversight by FDA, in particular, as an option to obtain more information on supply chains—information that is useful in tracing the source of economic adulteration. For example, one stakeholder suggested that the use of track-and-trace technology— such as using standard numerical identifiers on prescription drug packages—could facilitate FDA's oversight of the supply chain by making it easier for FDA and industry to trace adulterated ingredients back to the point of contamination. Under the new FDA Food Safety Modernization Act, the Secretary of HHS, acting through FDA, is directed to establish a system that will improve its ability to rapidly track and trace both domestic and imported foods. Similarly, the Food and Drug Administration Amendments Act of 2007 required FDA to develop a unique device identifier system to adequately identify a medical device through distribution and use. According to FDA officials, the agency expects to publish a proposed rule on the establishment of this system by the end of 2011.[23]

Many stakeholders also suggested that FDA increase its regulatory and enforcement actions to address economic adulteration. These stakeholders said that public health risk should be FDA's priority in taking such actions, but many also told us that FDA should pursue those who adulterate for economic gain, including in instances that may not have a large negative public health impact. For example, some stakeholders suggested building criminal cases against those who adulterate for economic gain and prosecuting them swiftly and visibly to help ensure that companies are complying with laws and regulations. In addition, these stakeholders said that, even when the adulteration has little health impact, such actions could help protect public health by deterring future instances, some of which may pose a significant health threat. Depending on the circumstances, such as the type of violation and product involved, a range of enforcement actions or penalties could be pursued. However, in February 2009, we reported that FDA has taken few actions in pursuing instances of economic fraud in seafood.[24] In that report, we found that FDA did not issue any regulatory letters to companies regarding seafood fraud from 2005 through 2008, and according to a senior FDA official, the agency had not taken any enforcement actions for seafood fraud since 2000.

Even with the challenges related to the disclosure of proprietary information, stakeholders also suggested that greater communication with industry could enhance FDA efforts to gather information on economic adulteration. One option for greater communication that several stakeholders identified was the creation of an information clearinghouse, through which companies could anonymously share information on adulterated ingredients with FDA and other companies. Stakeholders noted that the clearinghouse could enhance FDA's ability to disseminate information on adulterated products quickly, facilitate secure information sharing across industries, and enable FDA and industry to respond more rapidly to potential instances of adulteration. For example, they said that a clearinghouse could allow the sharing of information, such as information on market price fluctuations, environmental disasters, or other macroeconomic factors. In the view of these stakeholders, this type of information may help both industry and FDA better target their efforts to detect and prevent economic adulteration. One stakeholder said that such a clearinghouse was an opportunity for industry and FDA to share information from various sources in a central location, which would help them draw conclusions about the authenticity of ingredients or raw materials. This stakeholder suggested that if an information clearinghouse had existed prior to the heparin incident, it could have contained critical information—such as the sudden increase or decrease in the price of ingredients for food or drugs —to alert FDA and industry to the potential for adulteration.

One stakeholder noted that because some of the industries affected by economic adulteration are small, some companies might easily be identified by the information reported, even if they reported it anonymously. Consequently, some stakeholders suggested engaging a neutral third party to operate the information clearinghouse, thus helping to ensure that the information shared was free of specific company identifiers. FDA officials said that they are examining various ways to facilitate information sharing with industry and have discussed the idea of a clearinghouse, but they have no plans to develop one.

In addition to formal information sharing, some stakeholders suggested more informal interaction between industry and FDA. Stakeholders noted that increased dialogue could provide opportunities for FDA to communicate to industry its overall strategy on economic adulteration.

Some stakeholders told us that FDA's communication during adverse public health events was clear and timely but that at other times they were unsure what FDA was doing to address potential economic adulteration. Some stakeholders expressed a willingness to work with FDA on the issue but said that they need to better understand FDA's expectations of industry. For example, one stakeholder suggested a forum where FDA officials can talk to industry directly and engage in dialogue to clarify the agency's strategy.

Some stakeholders from food industry groups also said that they believe the recent passage of the FDA Food Safety Modernization Act provides new opportunities for both FDA and industry to address economic adulteration. One stakeholder noted that the new law may give FDA more opportunities to include economic adulteration in its inspection program. In addition, stakeholders told us that they believe the law provides a science- and risk-based approach for companies to verify their ingredient suppliers, including multiple ways of assuring the public and FDA that industry has processes in place to detect economic adulteration. Specifically, under the act, certain facilities are required to identify reasonably foreseeable hazards and to prepare written control plans that illustrate reasonable approaches to looking for intentional adulteration.

Lastly, one stakeholder said that FDA may need additional authority to require the drug industry to provide the agency with information critical to securing the medical product supply chain. Additional authority may include, for example, allowing FDA to require enhanced documentation from industry on its supply chains to increase transparency. In its comments on one of our recent reports, HHS also mentioned legislation previously under consideration by Congress that it believed would, if enacted, provide FDA with helpful tools to further secure the nation's drug supply chain.[25] For example, according to the agency, the proposed legislation would have provided FDA authority to require foreign and domestic drug manufacturers to implement quality systems and adopt plans to identify and mitigate hazards. In its comment letter, FDA said that such legislation could ensure that the agency can hold industry accountable for the security and integrity of its supply chains and quality control systems.

CONCLUSION

Economic adulteration is not a new problem. It can undermine confidence in the safety of the nation's food and medical products and have significant economic consequences for industry. The recent crises involving the contamination of pet food with melamine and the adulteration of heparin with oversulfated chondroitin sulfate showed that economic adulteration continues to be a problem and can have serious public health consequences. Senior FDA officials, including the Commissioner, have often spoken publicly about the threat posed by economic adulteration. However, FDA does not have a definition of economic adulteration. Without such a definition, when FDA detects adulteration, it is more difficult for the agency to make a distinction between economic adulteration and other forms of adulteration to guide the agency's thinking about how to be more proactive about this issue. In addition, FDA has not provided guidance to its centers and offices on how they should approach or address their economic adulteration efforts. This is not consistent with federal standards of internal control, which state that agencies should have documented policies and procedures in place to carry out management's directives. Some entities have undertaken efforts that specifically focus on economic adulteration, but they have not always communicated or coordinated their efforts with other FDA entities. Without such communication and coordination, in these times of economic uncertainty, FDA may not be making the best use of its scarce resources. As food and medical product supply chains become increasing global and complex, economic adulteration will continue to remain a threat.

RECOMMENDATIONS FOR EXECUTIVE ACTION

To enhance FDA's efforts to combat the economic adulteration of food and medical products, we recommend that the Commissioner of FDA take the following three actions:

- adopt a working definition of economic adulteration,
- provide written guidance to agency centers and offices on the means of addressing economic adulteration, and
- enhance communication and coordination of agency efforts on economic adulteration.

AGENCY COMMENTS AND OUR EVALUATION

We provided a draft of this report to HHS for review and comment. HHS neither agreed nor disagreed with our recommendations. In its comments, HHS stated that FDA views the term "economically motivated adulteration" as describing a subset of cases within the broader concept of adulteration, and believes that a holistic approach toward understanding and addressing adulteration generally is the best course forward. HHS also said that this approach will best serve the agency as it strives to protect the health and well-being of the American people by preventing, detecting, and taking appropriate responses to all adulterations of food and medical products. As we note in our report, however, agency entities have missed opportunities to communicate and coordinate efforts specifically directed at economic adulteration and identify potential public health risks. At the same time, FDA said that it recognizes the importance of sharing and leveraging information relevant to economically motivated adulteration and the utility of a mechanism for facilitating such sharing and collaboration at FDA. The department provided additional information in its written comments on planned actions of FDA's

Working Group on Economically Motivated Adulteration that are consistent with two of the three recommendations we made in our draft report. The additional comments related to our recommendations that FDA adopt a working definition of economic adulteration and enhance communication and coordination of agency efforts on economic adulteration are as follows:

- *Adopt a working definition of economic adulteration.* HHS stated that the Working Group on Economically Motivated Adulteration will use the working definition of economically motivated adulteration that FDA proposed at its May 2009 public meeting on the topic.
- *Enhance communication and coordination of agency efforts on economic adulteration.* HHS stated that FDA expects the efforts of the working group will result in enhanced collaboration and communication at FDA on ways to approach and address situations of economically motivated adulteration.

We have included this additional information in our report.

HHS also provided technical comments, which we incorporated as appropriate.

Lisa Shames
Director, Natural Resources and Environment

Marcia Crosse
Director, Health Care

APPENDIX I: OBJECTIVES, SCOPE, AND METHODOLOGY

This report examines (1) the approaches the Food and Drug Administration (FDA) uses to detect and prevent economic adulteration of food and medical products, and (2) the challenges, if any, FDA faces in detecting and preventing economic adulteration and stakeholder views on options for FDA to enhance its efforts to address economic adulteration. For this report, we define economic adulteration as "the fraudulent, intentional substitution or addition of a substance in a product for the purpose of increasing the apparent value of the product or reducing the cost of its production, i.e., for economic gain. [It] includes dilution of products with increased quantities of an already-present substance (e.g., increasing inactive ingredients of a drug with a resulting reduction in strength of the finished product, or watering down of juice) to the extent that such dilution poses a known or possible health risk to consumers, as well as the addition or substitution of substances in order to mask dilution." Our definition of economic adulteration is the same as the working definition of "economically motivated adulteration" that FDA developed for a May 2009 public meeting to raise awareness and solicit input on the topic. We did not include counterfeiting of a finished product because counterfeiting concerns the unauthorized use of intellectual property rights.

To determine the approaches FDA uses to detect and prevent economic adulteration of food and medical products, we interviewed officials from the five FDA centers responsible for food and medical products, including the Center for Food Safety and Applied Nutrition, the Center for Veterinary Medicine, the Center for Drug Evaluation and Research, the Center for Devices and Radiological Health, and the Center for Biologics Evaluation and Research, as well as FDA's Office of Regulatory Affairs, Office of International Programs, and Office of the Commissioner. We also interviewed former FDA officials and representatives of organizations that have been assisting FDA in its efforts to detect and prevent economic adulteration, including the United States Pharmacopeia, the University of Minnesota's National Center for Food Protection and Defense, and New Mexico State University's Center for Animal Health, Food Safety and BioSecurity. We reviewed relevant FDA documents, including regulations, compliance manuals and inspection guides, sampling surveillance results, statements and presentations by agency officials, a contract to fund a research project at the National Center for Food Protection and Defense, and communications with industry and the public. We also reviewed published information from FDA, including its *Strategic Priorities 2011- 2015* report, its *Pathway to Global Product Safety and Quality* report, and *Federal Register* notices. We also reviewed previous GAO reports and recommendations on FDA's oversight of food and medical products, as well as the agency's strategic planning efforts. We compared FDA's efforts to address economic adulteration with federal standards for internal control.[26]

To determine the challenges FDA faces in detecting and preventing economic adulteration, we interviewed and obtained the views of FDA officials and stakeholders about the challenges the agency faces in addressing economic adulteration. Stakeholders included members of academia and representatives of industry and consumer groups who made presentations at FDA's May 2009 meeting on economically motivated adulteration, as well as former FDA officials who were involved in agency efforts that led to that meeting. We also interviewed and obtained the views of the stakeholders on options for FDA to enhance its efforts to address economic adulteration. The views of these stakeholders are not representative of and cannot be generalized to all stakeholders. In addition, we reviewed FDA and stakeholder documents related to challenges and options, as well as portions of the FDA Food Safety Modernization Act.

We conducted this performance audit from September 2010 to October 2011 in accordance with generally accepted government auditing standards. Those standards require that we plan and perform the audit to obtain sufficient, appropriate evidence to provide a reasonable basis for our findings and conclusions based on our audit objectives. We believe that the evidence obtained provides a reasonable basis for our findings and conclusions based on our audit objectives.

End Notes

[1] According to FDA officials, the imported vegetable protein products also contained cyanuric acid, a melamine-related compound. Melamine and cyanuric acid individually are relatively nontoxic. However, when combined, they produce crystals, which can lead to kidney failure. For purposes of this report, we use the term melamine to mean melamine and its related analogs (e.g., melamine and cyanuric acid).

[2] Biologics are generally materials—such as vaccines—derived from living sources, such as humans, animals, and microorganisms, as well as materials produced by biotechnology methods. See 42 U.S.C. § 262(i); 21 C.F.R. § 600.3(h).

[3] GAO, *High-Risk Series: An Update*, GAO-1 1-278 (Washington, D.C.: February 2011).

[4] GAO, *High-Risk Series: An Update*, GAO-07-310 (Washington, D.C.: January 2007); and *High-Risk Series: An Update*, GAO-09-271 (Washington, D.C.: January 2009).

[5] See 74 Fed. Reg. 15498 (Apr. 6, 2009).

[6] For the purposes of this report, we are using the term "economic adulteration" to mean "economically motivated adulteration."

[7] The United States Department of Agriculture is responsible for ensuring the safety of meat and poultry products.

[8] According to FDA's Office of Chief Counsel, formal enforcement actions available to FDA include initiating a seizure of an adulterated product, seeking an injunction to stop a company from engaging in certain behavior, or referring a firm for criminal prosecution. FDA can also issue warning letters, which are intended to prompt companies to voluntarily correct violations of regulatory significance.

[9] GAO, *Food and Drug Administration: FDA Faces Challenges Meeting Its Growing Medical Product Responsibilities and Should Develop Complete Estimates of Its Resource Needs,* GAO-09-581 (Washington, D.C.: June 19, 2009).

[10] GAO, *Food and Drug Administration: Opportunities Exist to Better Address Management Challenges*, GAO-10-279 (Washington, D.C.: Feb. 19, 2010).

[11] GAO, *Food and Drug Administration: Overseas Offices Have Taken Steps to Help Ensure Import Safety, but More Long-Term Planning Is Needed*, GAO-1 0-960 (Washington, D.C.: Sept. 30, 2010).

[12] The official drug compendium of the United States is the *United States Pharmacopeia/National Formulary*, published by the United States Pharmacopeial Convention, a non-governmental, standard-setting authority.

[13] For more information on PREDICT, see *Food Safety: FDA Has Begun to Take Action to Address Weaknesses in Food Safety Research, but Gaps Remain*, GAO-10-182R (Washington, D.C.: Apr. 23, 2010).

[14] API refers to any component that is intended to provide pharmacological activity or other direct effect in the diagnosis, cure, mitigation, treatment, or prevention of disease.

[15] These standards are in the form of monographs—written standards that describe a substance or product, including information on tests needed to ensure that the substance is of the appropriate strength, quality, and purity.

[16] The National Center for Food Protection and Defense is a research consortium that addresses the vulnerability of the nation's food system to attack through intentional contamination with biological or chemical agents.

[17] While CVM has not undertaken any efforts exclusively targeting economic adulteration, in its written comments on our draft report, HHS noted that CVM is participating in an FDA Monograph Modernization Task Group to help USP prioritize monographs in need of modernization, which includes reduction of economic adulteration.

[18] GAO-1 0-279.

[19] Department of Health and Human Services, U.S. Food and Drug Administration, *Pathway to Global Product Safety and Quality* (Washington, D.C.: July 2011).

[20] GAO, *Standards for Internal Control in the Federal Government,* GAO/AIMD-00-21 .3.1 (Washington, D.C.: November 1999).

[21] GAO, *Medical Devices: Shortcomings in FDA's Premarket Review, Postmarket Surveillance, and Inspections of Device Manufacturing Establishments*, GAO-09-370T (Washington, D.C.: June 18, 2009).

[22] GAO-09-581, and GAO, *Food and Drug Administration: Effect of User Fees on Drug Approval Times, Withdrawals, and Other Agency Activities*, GAO-02-958 (Washington, D.C.: Sept. 17, 2002).

[23] GAO, *Medical Devices: FDA Should Enhance Its Oversight of Recalls*, GAO-1 1-468 (Washington, D.C.: June 14, 2011).

[24] GAO, *Seafood Fraud: FDA Program Changes and Better Collaboration among Key Federal Agencies Could Improve Prevention and Detection*, GAO-09-258 (Washington, D.C.: Feb. 19, 2009).

[25] GAO-1 1-95.

[26] GAO, *Standards for Internal Control in the Federal Government,* GAO/AIMD-00-2 1.3.1 (Washington, D.C.: November 1999).

INDEX

A

B

C

D

E

F

G

H

I

J

K

L

M

N

O

P

Q

R

S

T